ALH1

CW00375728

SATELLITE OPERATIONS
Systems Approach to Design and Control

THE ELLIS HORWOOD LIBRARY OF SPACE SCIENCE AND SPACE TECHNOLOGY

SERIES IN SPACE TECHNOLOGY

Series Editor: D. R. SLOGGETT, Software Sciences Limited, Farnborough, Hampshire
METALLURGICAL ASSESSMENT OF SPACECRAFT PARTS AND MATERIALS
BARRIE D. DUNN, Head of Metallic Materials and Processes Section, ESA-ESTEC
OPTICAL INSTRUMENTS IN REMOTE SENSING
B. PEASE
AUTOMATION IN SPACE: Vol. 1: Expert Systems in Space; Vol. 2 Robots in Space
DAVID R. SLOGGETT, Principal Consultant, Software Sciences Limited, Farnborough, Hampshire
SATELLITE DATA: Processing, Archiving and Dissemination
Vol. I: Applications and Infrastructure
Vol. II: Functions, Operational Principles and Design
DAVID R. SLOGGETT, Principal Consultant, Software Sciences Limited, Farnborough, Hampshire
PHOTOGRAPHIC PROCESSING OF REMOTELY SENSED DATA
GRANT THOMSON, Hunting Technical Services, Borehamwood, Hertfordshire
DIGITAL CAROGRAPHY FROM SPACE
JONATHAN M. WILLIAMS, SERCO Limited, Sunbury on Thames

SERIES IN SPACE ENGINEERING

SATELLITE OPERATIONS: Systems Approach to Design and Control
JOHN T. GARNER and MALCOLM JONES, ESTEC, The Netherlands

SERIES IN SPACE LIFE SCIENCES

MARS AND THE DEVELOPMENT OF LIFE
Editor: ANDERS HANSSON, Deputy Director of the Institute for Space Biomedicine, University of Sheffield, UK

SERIES IN REMOTE SENSING

APPLICATIONS OF WEATHER RADAR SYSTEMS:
A Guide to Uses of Radar Data in Meteorology and Hydrology
CHRISTOPHER G. COLLIER, Assistant Director (Nowcasting & Satellite Applications), Meteorological Office, Bracknell

SATELLITE OPERATIONS
Systems Approach to Design and Control

JOHN T. GARNER C Eng MIEE
Principal Ground Support Engineer,
Communications Satellite Programmes,
ESA-ESTEC, Noordwijk, The Netherlands

MALCOLM JONES M I Elec IE
Senior Engineer, Serco Space Ltd
based at ESTEC, Noordwijk, The Netherlands

ELLIS HORWOOD
NEW YORK LONDON TORONTO SYDNEY TOKYO SINGAPORE

First published in 1990 by
ELLIS HORWOOD LIMITED
Market Cross House, Cooper Street,
Chichester, West Sussex, PO19 1EB, England

A division of
Simon & Schuster International Group
A Paramount Communications Company

Typeset in Times by Ellis Horwood Limited
Printed and bound in Great Britain
by Hartnolls Limited, Bodmin, Cornwall

British Library Cataloguing in Publication Data

Garner, John T.
Satellite operations.
1. Communications satellites. Systems engineering
I. Title II. Jones, Malcolm
621.3825
ISBN 0–13–791351–6

Library of Congress Cataloging-in-Publication Data

Garner, John T., 1943–
Satellite operations: systems approach to design and control /
John T. Garner, Malcolm Jones.
p. cm. — (Ellis Horwood library of space science and space
technology. Series in space technology)
ISBN 0–13–791351–6
1. Artificial satellites design and construction. I. Jones,
Malcolm, 1935– . II. Title. III. Series.
TL796.G28   1990
629.46–dc20   90–35292
                                                          CIP

Contents

Preface

In the early days of space exploration satellite operations were undertaken to meet the needs of research aims. These pursuits are still undertaken; however satellites can now be placed in near Earth orbits for purposes that embrace commercial aspects. Communication satellites are an example of such an application within the aerospace industry.

The authors are of the opinion that cost-effectiveness is not being adequately considered within the developing aerospace industry. In their experience, too much duplication of developments is being undertaken. One of the contributing factors is the confusion which is being caused by the use of dual terminology for aerospace system elements, particularly with system items which are not launched and remain on Earth. A fully operational aerospace system does have components which do not fly but nevertheless can be vital for ensuring safe operations to be conducted and maintained.

Some rationalization of spacecraft operations is taking place but to enable efficient programmes to be undertaken, some guidelines need to be aired. This is an objective of the book which takes satellite operations as an example of how engineering efficiency can be improved.

In the opinion of the authors, a complete systems approach to aerospace system design and development should actively consider the pre-launch scenario. This methodology is being practised to some extent but more attention to its implementation should be established. Consequently, the pre-launch activities should be a baseline to ensure that system applications can meet post-launch requirements.

The book describes the facilities and methodologies which are necessary for the efficient construction and operation of unmanned satellites which orbit the Earth. A communication system is taken as the prime example for presenting a fundamental approach to the application of systems engineering technologies for the exploitation of space. The book describes in a general manner how the various units and subsystems comprising a communication spacecraft are integrated and operated. There is no attempt to explain in detail the operation of individual spacecraft subsystems, although system level aspects are covered. In-flight operations are also dealt with in an expansive but undetailed manner. Both pre- and post-launch satellite operational disciplines and methodologies are compared, particularly in the satellite control domain. The book presents a case for consideration in the satellite operations field where a rationalization process could harmonize pre- and post-launch activities. With this approach a satellite launch could be less of a step function in an aerospace programme.

The subject matter of the book is treated with a systems engineering methodology combined with a fundamental approach. Therefore, other publications can and do deal with some subjects of this book in more technical detail, but perhaps in a somewhat insular manner.

The combined experiences of the authors span a period greater than twenty years in the aerospace industry. During this time they have had direct responsibilities for providing and ensuring that spacecraft electrical support equipment has been constantly available to aid satellite operations before launch; these pre-launch activities commence with spacecraft construction and terminate at launch vehicle lift-off. They have provided such support to both scientific and application spacecraft programmes, embracing over fifteen launch campaigns. Their prime experiences are centred on many of the communication satellites that have been constructed in Europe, and supporting roles have encompassed aerospace programmes which have been undertaken within and outside the auspices of the European Space Agency (ESA). Consequently, the authors present a European point of view coupled with their practical experiences. This is particularly evident in the presentation of the computer software designated the European test and operations language (ETOL) which has been employed for pre-launch spacecraft operation. The authors appreciate that other languages exist for this task, but their prime aim was to be practical and use an example of computer software that has been actively used for satellite pre- and post-launch operations tasks.

The authors have organized and presented training courses for members of the European aerospace industry and some development organizations of the Third World. They have had several articles published by the technical press, and have extracted relevant details from these articles for inclusion in the book. The book is the result of their practical experiences which is hoped can demonstrate to the reader that spacecraft technologies and operations are engineering disciplines and not a form of mystic.

The authors consider that this book will be of value to personnel actively engaged in the international aerospace industry, including junior engineers, personnel entering the industry and experienced engineers who have specialized in other fields. Furthermore, they anticipate that the book will be useful for administrative and financial controllers of spacecraft programmes, besides the technical managers, providing these specialists with background information regarding overall system aspects of aerospace systems. The authors also hope that the book will be of interest to professionals who are engaged with engineering system design and operations that are outside the aerospace industry.

The authors would like to thank colleagues for their assistance which may in some cases have been indirect and unknown to them. In addition thanks are extended to Miss Rita Dron for her expertise with her word processing facilities, Mr Norman Thomas for his editorial support, and Mr Pim Muijzert who produced so many drawings. Their efforts made an invaluable contribution to the production of the manuscript.

The views expressed in the book are those of the authors and are not necessarily representative of their employers — ESA and SERCo Space Ltd.

Abbreviations

ABM	Apogee Boost Motor
AIT	Assembly, Integration, and Test
AIV	Assembly, Integration, and Verification
AOCS	Attitude and Orbit Control System
AOS	Acquisition Of Signal
ASW	Address and Synchronization Word
CM	Communication Module
CSMF	Communications Satellite Monitoring Facility
DRS	Data Relay System
ECS	European Communication Satellite
EGSE	Electrical Ground Support Equipment
ELV	Expendable Launch Vehicle
EM	Engineering Model
EOL	End Of Life
ESA	European Space Agency
ETOL	European Test and Operations Language
EURECA	EUreopean REtrievable CArrier
EUTELSAT	EUropean TELecommunication SATellite organization
FM	Flight Module
FOP	Flight Operations Plan
GEO	Geostationary Earth Orbit
GSA	Ground Station Adaptor
GSE	Ground Support Equipment
GTO	Geostationary Transfer Orbit
INMARSAT	INternational MARitime SATellite organization
IOT	In-Orbit Testing
IST	Integrated Systems Test
LAN	Local Area Network
LEO	Launch and Early Orbit
LEOP	Launch and Early Orbit Phase

MARECS	MARitime European Communications Satellite
MGSE	Mechanical Ground Support Equipment
MMI	Man/Machine Interface
MTGP	Monitor Table Generation Program
MTP	Master Test Processor
NASA	National Aeronautics and Space Administration (USA)
NRZ-L	Non-Return to Zero-Level
OBDH	On-Board Data Handling
OCC	Operations and Control Centre
OCOE	Overall CheckOut Equipment
ORA	OCOE Remote Adaptor
OTS	Orbital Test Satellite
PCM	Pulse Code Modulation
PFM	Prototype Flight Model
P/L	Payload
PSK	Phase Shift Keying
PSS	Power Supply Subsystem
QM	Qualification Model
RF	Radio Frequency
Rx	Receiver
SCOE	Specific CheckOut Equipment
SCS	Spacecraft Control System
SM	Service Module
SPL	Split Phase Level
TAB	Timing and Acquisition Bits
TC	Telecommand
TCC	Test Conductor Console
TDM	Time Division Multiplex
TM	Telemetry
TMS	Test and Monitor Station
TTC	Telemetry, Tracking and Command
Tx	Transmitter
V	Voltage
VDU	Visual Display Unit

1

Aerospace systems:
structure and philosophy

1.1 SYSTEM EVOLUTION

Some features of the aerospace industry's modern engineering beginnings can be based upon exploits of the Wright Brothers in the early 1900s. The development of their heavier than air flying machine, the aeroplane, has been fundamental to the development of today's aerospace industry. At the beginning of the 20th century the idea of being able to fly over and between continents was not envisaged by the general public. Indeed in the 1950s the idea of man being able to land upon and explore the Moon was still a subject of science fiction. Now aeroplanes can fly around the Earth in hours, and space travellers have visited the Moon. Some members of mankind have lived on a spacecraft for days, weeks, and months in the non-terrestrial environment known as space. These endeavours form the basis for continued developments within the many facets of the aerospace industry which they have nurtured.

The term 'aerospace' relates to activities which take place above the surface of the Earth, both within and external to the Earth's gravitational forces and atmosphere. For the purposes of this book, the term aerospace relates directly to the environment in which satellites orbit the Earth, although more expansive aerospace subjects are covered to some extent; descriptions concerning launch vehicles and rockets fall into this expansive category.

The development of aerospace systems will continue through the next decade, into the next century and on into the more distant future. New and extended developments will certainly be undertaken in some fields. In other areas, developments will be necessary only in special cases. Indeed, a consolidation and rationalization of established and developing technologies could be fruitful. The garnering of rewards that can follow earlier development activities should be possible, and in the commercial world imperative. Suitable engineering designs will allow for re-use of established technology, providing a desirable exploitation of space. The prime purpose of this book is to demonstrate that the technology associated with the operation of satellites can be re-used and employed for activities that take place before and after they are launched into space. This should enable developments to be beneficial from both cost-effective and technical performance standpoints. The technologies which have been developed for spacecraft operations could also be adapted for use by systems which are not part of the aerospace industry.

1.2 SYSTEM OVERVIEW

An aerospace system comprises two prime parts which are designated the Earth segment and the space segment. The segments need to operate in unison, and they can each be composed of more than one element. Within the Earth segment, some elements which perform identical tasks can be termed differently. This difference in terms is primarily dependent upon which phase of the aerospace programme is being conducted. An objective of the book is to give a concise definition of what elements constitute an Earth segment. This clarification leads into another objective which is to consider the possibilities of re-using aerospace system components to the maximum extent. Such re-use can be applicable for equipment utilization before and after spacecraft launches. Furthermore, applications of technology re-use between spacecraft programmes can be achieved if consideration is given to this possibility in the early stages of aerospace system designs. A significant factor in achieving the re-use goal is concentrated around the structure of an executive control centre with methods and implementation of delegatory powers being important criteria.

The launch of a spacecraft from the Earth should be considered as a milestone in the construction of an aerospace system and not a start of the process (Fig. 1.1). Applications satellites which are in orbits near to the Earth are exactly the same before and after their launch; it is their distance from Earth-based control and user facilities, and the space environment that dominate the need for variations in operational methods. This book adopts a systematic approach in reviewing satellite operations. Therefore, pre-launch methodologies form the baseline for analysis, with post-launch requirements being a dominant factor for operational considerations.

The post-launch Earth segment is employed for satellite control and satellite use. The pre-launch Earth segment is used to practise operations and test satellite performance. This verification process, if adequately conducted, may require test facilities which are more performant than the satellite under test. Satellite operations, both pre- and post-launch, require Earth segment support facilities continuously. This requirement will commence during engineering development and continue until the satellite ceases to be operational in orbit. Therefore, a modern operational aerospace system requires support from the beginning to the end of a programme.

1.2.1 Applications

The aerospace industry which had its exploitive beginnings with aeroplanes that fly above the Earth's surface has expanded dramatically since the launching of exploratory probes into space. These explorations have prompted the development of techniques which can be beneficial for space exploitation endeavours. Technological

Fig. 1.1 — Ariane-VI launch vehicle, a few seconds after engine ignition (courtesy of ESA).

developments have enabled aerospace systems to be developed and utilized to improve life on Earth.

Satellites can aid the management of the planet's natural assets and the provision of information for weather forecasting. Satellites that remain at a fixed position above the Earth (in geostationary orbit) have enabled communication links to be established across wide areas of the Earth (Fig. 1.2). These links facilitate the establishment of communications between fixed and mobile terminals, thus making a valuable contribution to the quality of life on Earth.

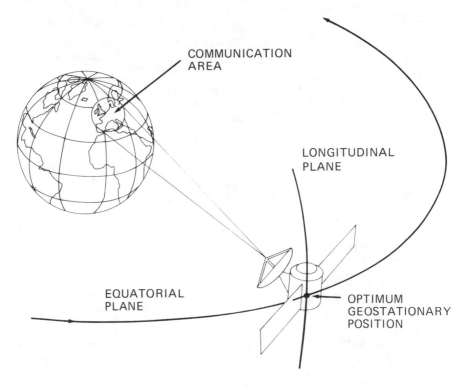

Fig. 1.2 — Geostationary position. Pictorial view of a communication satellite in geostationary Earth orbit that enables links to be established over Europe (courtesy of ESA).

These endeavours will be truly beneficial only if they are efficient; this can be interpreted as not being wasteful. Certainly, spacecraft can efficiently utilize and exploit geostationary Earth orbits whilst they are operational and providing a service. After these spacecraft cease to function as useful elements of an aerospace system then they can become debris that chokes geostationary orbital positions. The direct disadvantage of this is that the position cannot be re-used until the spacecraft debris no longer exists. With increasing manned flights into space such man-made debris, whether geostationary or in other Earth orbits, can become a danger or at least a problem for the construction and operation of developing aerospace systems. This growing operational constraint deserves attention but is somewhat outside the scope of this book. Nevertheless, it can and should be addressed during the design phases of an aerospace system.

The benefits of space exploitation can be indirectly harvested. With the advent of space stations, an acceptable living environment for human beings will necessitate not waste disposal but waste recycling. Perhaps the planet Earth can benefit from such exploitive endeavours. If solutions to problems that exist outside the close environment can be reworked and implemented on the Earth then benefits to life on the planet should result. Indeed, the exploitation of space if truly beneficial will mean 'use to advantage', but with no disadvantages or unacceptable side effects.

The exploitation of space has passed from its beginnings when it was a technological development that resulted in the launch and low Earth orbit of the USSR's Sputnik in 1959. Since then the industry has expanded both from technological and international standpoints, bringing closer together the users and the providers of aerospace systems.

The development of aerospace systems, because of the nature of the effort entailed, has required the expert assistance of many people from the various areas of industrial and academic life. Certainly, scientific and commercial endeavours have brought together a large variety of engineering and management techniques, and people of different nationalities.

The endeavours of space scientists and astronomers, together with their engineering colleagues, have resulted in the possibility for the commercial exploitation of space. These exploitation endeavours need managing correctly if the benefits are to be all embracing rather than insular. A review of development activities past, present and future needs to be made periodically. This will allow technical standards to be agreed and working practices rationalized.

Initially, exploratory endeavours were formulated to meet visionary scientific aims. At that time, exploration and exploitation activities were closely merged. Now, with the advance of technologies that have been prompted by space exploitation and exploration, these pursuits have separated more clearly into scientific and commercially motivated programmes. Space exploration and exploitation can and should continue in a coordinated manner based on the experiences gained. A systematic approach will be comprehensive and methodical, allowing the industry to expand if managed efficiently.

In the not too distant future the general public should be able to reap more fully and directly the benefits of the aerospace industry, which they have indirectly fostered, by providing funds to satisfy the aspirations of scientists and politicians. Improved communications, including TV reception, and more reliable weather forecasts are some of the rewards that are more generally directly available. The achievement of satisfactory benefits will depend heavily upon the implementation of cost effective solutions. Such solutions will enable aerospace investments to be adequately rewarded. These will be dependent upon the management of the industry. A marketplace will provide what customers require at a price. The price that a customer pays in a good business environment will give satisfaction to all.

1.2.2 Management
The aerospace industry is very forward-looking in aspects which are motivated by exploration and the development of new techniques. The application of the asso-

ciated technologies, including those which are established or under development, needs to be managed with skill. Besides the techniques of science and engineering, there are the techniques of management. Management techniques need to be applied with diligence if facilities and resources are to be employed in a manner that enables safe and efficient aerospace systems to be developed, produced, and operated to a time-scale that meets user requirements.

The most fundamental aspect of any system design should encompass the ultimate users' requirements as being of prime importance. System engineering principles should then embrace cost effectiveness to give the best engineered solution.

The construction and operation of the earliest spacecraft with scientific goals brought closely together a variety of engineering disciplines to produce an operable aerospace system. During construction, electronic, mechanical and thermal experts were drawn together. Engineering expertise was aligned with specialists who undertook tasks associated with spacecraft flight predictions. The combination of all these experts continued after the spacecraft had been launched so that the needs of the spacecraft users — the scientists — could be met. The integration of the multiple disciplines embraced by such an aerospace system can be complex from technological and managerial aspects. The efforts which produced the first operational satellites embraced changes to established working practices, and which continue owing to the expansion of techniques which are required for the operation of advanced spacecraft.

The construction process of a spacecraft can require the use of testing facilities which are at different geographical locations. Certainly, the location of launch sites can involve transportation of spacecraft from their construction sites to where the launch facilities are located. Testing activities are normally undertaken at launch sites, which encompass verifications of spacecraft — launch vehicle interfaces at least. All the pre-launch spacecraft operations need activities at the different geographical locations being performed by various teams of engineers that manage spacecraft and test facility activities. Indeed, aerospace systems are constructed and operated by teams comprising a comprehensive range of abilities.

A comprehensive programme will require that sufficient resources of adequate quality are available from the initiation of a space programme to its cessation. A programme lasts for many years and can produce management problems if the complete requirements are not addressed at the initial stages of definition and design. Similarly, requirements should be reviewed at regular intervals throughout the lifespan of the system. These reviews should cover all aspects of resource provisions to ensure that unnecessary duplications which would reduce efficiency are not produced. The scope of aerospace programmes is expanding, hence a macro-management approach is required to coordinate all the tasks necessary to produce an efficient operational aerospace system.

1.3   A SYSTEMATIC APPROACH

A complete aerospace system will comprise Earth-based elements and space elements that have been launched from the Earth. The design and operation of one element group should always consider the requirements of the other. This will be necessary if the aerospace system is to be produced and operated efficiently. The spacecraft will need to be supported by facilities that remain on Earth for both pre- and post-launch activities. An objective of this book is to enable clarifications to be made regarding these Earth-based facilities so that re-use can be more easily analyzed and implemented.

A spacecraft is part of an aerospace system which operates under the auspices of a controlling element. Control is effective during the construction of the spacecraft, its launch, and finally its operation in space. Indeed, throughout all phases of an aerospace programme a spacecraft will be subjected to a variety of operational environments. A space element (manned, unmanned, a deep probe, or in Earth orbit) will be required to remain in contact with an executive controller. This being the case, then a completely independent space element does not exist and most likely a design never will be implemented for complete and permanent autonomous operation.

The term 'operations', when relating to spacecraft of a fully operational aerospace system, needs to be reviewed. When fully operational, a spacecraft will be controlled to meet users' requirements. Therefore, for spacecraft post-launch activities, operations are subdivided into control and use. Control activities will be aligned to keeping the spacecraft at the status required by users, and maintaining the orbital position to satisfy their needs. Satisfactory control will enable users to employ the spacecraft to suit their purpose. A communication satellite which operates to provide links between ultimate users is an example of how a space segment can be applied and employed.

Pre-launch spacecraft operations will be associated with construction and design verification, subdividing operations into control and test.

1.4 SYSTEM CONFIGURATIONS

Aerospace systems comprise two prime elements which as stated previously are termed the space segment and the Earth segment.

A space segment can consist of a scientific or application type of spacecraft. A scientific spacecraft is one that is engaged upon a scientific mission of exploration, whereas an application spacecraft is of a kind that allows its orbital paths to be used for more commercially motivated purposes. The former will carry equipment for scientific experimental purposes; the latter provides services to the human community.

A satellite is a type of spacecraft that travels in a regular, clearly defined orbit around the centre of gravity of another celestial body. A spacecraft engaged upon a deep space mission and which is not travelling in a path around another body in a regular period is not a satellite in the exact definition, but is clearly a spacecraft. The exploration and exploitation of space embrace satellites which orbit the Earth, deep space probes, and manned or unmanned spacecraft, dependent upon the purpose of the missions. Thus, a spacecraft is a vehicle which is deployed from the strong gravitational attractions of the Earth with men and/or technological packages on board.

It is within the Earth segment where the user and control facilities of an aerospace system are normally located. The control capability which coordinates spacecraft operations with user requirements is often termed the 'ground segment' rather than 'ground support facilities' which describes its true function. For spacecraft control to be effective, there are requirements for liaisons and interfaces between the segments. Therefore, some form of a communication system will always be an integral part of an aerospace system so that messages can be exchanged between segment elements for operations and control.

Before dealing with more detailed aspects of systems that embrace components which operate in space and are controlled from the Earth, consideration needs to be given to some basic facts.

1.4.1 System components

An efficient system will have a structured architecture which is well-defined, comprising an ordered hierarchy of components that are engineered in a modular manner. A component may be another system which, within the hierarchy, can be subjected to control by an executive. An executive or master within a system is that component which has overall control, and can delegate control to other components. However, the executive should retain the master role and have the ability to coordinate and override its delegatory decisions, giving a control centre methodology.

A component which operates under an executive, either directly or indirectly, is termed a subsystem. Subsystems are elements that perform defined functions. Thus, a variety of subsystems operate in unison to perform system tasks. A subsystem is composed of items that are known as equipments or units. In an efficient system, the components described should be modules. Efficient modular components will perform a specific function and have defined external interfaces. Thus, with efficiently designed modular components, replacements and re-use can be readily achieved and implemented if interfaces and protocols remain unchanged.

With this understanding, a complete or overall system can be defined as end-to-end, which is regarded as input and output. Furthermore, top-down or bottom/up definitions are also applicable, these terms being relevant to a layered system hierarchy.

The performance of complete aerospace systems will be measured by their ultimate users. In the case of scientific probes, the ultimate users can be scientists who operate equipment which analyzes the results of their probing instruments on board a spacecraft. Indeed, efficient systems will operate to meet the exact requirements of the users. Complete systems consisting of both space and Earth segments are shown in simplistic form in Fig. 1.3. This figure depicts two examples of systems having a space segment. One example portrays a space explorer (a scientist) located on the Earth as the ultimate user of an aerospace system. The other example shows ultimate users as customers of a communication system using telephones.

1.4.2 Communication links

For both the systems shown in Fig. 1.3, the communication links between the Earth and space segments have commonalities. Both the scientific and communication spacecraft require control links to their ground support facilities. These control links are duplex (bidirectional) links which have similar basic requirements for both types of spacecraft. The links provide the remote control facilities for the transmission of

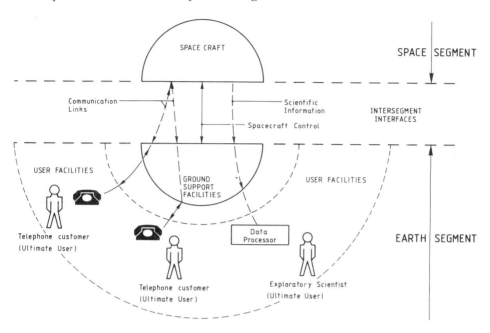

Fig. 1.3 — Complete system. The concepts of inter-segment links for spacecraft control and for ultimate users are shown.

orders to and the reception of data from a spacecraft. A duplex link is also employed to enable the flight path of the spacecraft to be monitored and controlled. Ground support facilities must enable these monitor and control functions to be accomplished.

Other communication links between ground support facilities from and to the scientific and the application spacecraft are not common. The scientific spacecraft uses a data link to relay the results from scientific observations. This communication link will be a simplex (unidirectional) link from the spacecraft to Earth. A communication spacecraft, on the other hand, does not have a user simplex link, but has user duplex links which enable it to fulfil its intended role in the complete communication system of relaying information between users.

Before continuing with this discourse, it is appropriate that a brief description is given for the telemetry and telecommand processes.

Telemetry: Information that has been gathered by the scientific spacecraft is encoded and transmitted from the spacecraft to the Earth. This information will require to be decoded and processed so that it can readily be employed by the user.

For both scientific and application endeavours, information concerning the condition and status of the spacecraft will be telemetered. This information, which is often termed housekeeping telemetry, also requires to be processed to enable it to be utilized efficiently for spacecraft control purposes.

Telecommands: Orders that are encoded and transmitted from the Earth to the spacecraft. These orders should configure and operate the spacecraft to the users' requirements. Such capabilities are necessary for spacecraft engaged upon exploration or exploitation missions.

The space segment for a scientist could be a single spacecraft which has been launched upon a probing mission. Control liaisons with the spacecraft will incorporate telemetry and telecommand signals that modulate radio frequency (RF) signals. The scientist will need to know the flight path and position of the spacecraft. This will necessitate another inter-segment interface associated with spacecraft control. These operations follow the path and determine the position of in-flight spacecraft are termed tracking operations. The post-launch position of a spacecraft can be followed and controlled by RF techniques. There are several methodologies which can be employed to perform the tracking functions. These take account of the fact that RF signals take time to travel from a transmitter to a receiver. It is this period, proportional to distance, which can be used to track and determine a spacecraft's orbital position.

As previously mentioned, the second overall system depicted in Fig. 1.3 is a system that has been manufactured for the technical exploitation of space. In this case, technologists and engineers have set up a system that provides services to the public. This example takes the ultimate users to be customers engaging a communication system. The space segment under these circumstances is a spacecraft which allows radio frequencies to relay information between the system users. The information transfers between customers can enable telephone conversations to take place, besides more technical interactions between users' computers. The customer can also be provided with a service that allows the viewing of television programmes. This type of spacecraft will be also subjected to control from the Earth. The tracking function can be performed with the RF signals that provide the communication links for spacecraft control by telemetry and telecommand. All of these control operations are carried out under the authority of the communication service supplier who will undertake the necessary actions to ensure that the spacecraft operates to meet the customers' needs.

1.4.3   System comparisons
The requirements for a complete aerospace system should be defined by or with the cooperation of the user who may be a scientist or an engineer. Astronomers and research scientists can be members of a scientific exploration category, whereas engineers concerned with space exploitation can be considered as members of a technical application category. The technical application engineers may not be the ultimate users of their system, but they can and should represent their customers who will be using the system. The exploitation and exploration of space can necessitate different approaches to the criteria associated with efficient system designs. Space explorations, such as the mid 1980s space probe associated with Halley's Comet, for example, are not dominated by costs alone. For space explorations, reliability, technical efficiency, and operational timescale requirements which define a launch date can dominate designs. Nevertheless, financial cost effectiveness should not be

ignored, but financial costs may be high to satisfy the other criteria. A re-use of ideas and designs from engineering developments associated with space exploration endeavours can and should be taken into account for the commercial aspects of space exploitation.

Application satellites can be considered as being concerned with the exploitation of space. Some types of these can be considered to be in an early development stage, whilst others are really in a more-definitive commercial field. Communication satellites are within this latter category, even though there are still significant development exercises going on. These will continue as the requirements of space exploration and exploitation evolve. Communication systems will always be a prime factor for the production and operation of efficient and safe aerospace systems.

Aerospace systems should be defined in the end-to-end system input and output context. There will be components between the two ends which can have prime, executive, major and minor roles to play in the system hierarchy. Provided that engineering modularity has been correctly implemented, operations should be efficiently performed. An efficient system can have different modes of operation to promote safety. These can include a spacecraft taking autonomous actions prompted by dangerous situations to achieve a safe condition. These recoveries may interfere with user access, but full services should be resumed when the fully operational status is reestablished.

The end-to-end context for a scientific probe will be the exploratory scientist at one extreme and the spacecraft at the other extreme. For the technical exploitation example, the configuration can have a user or a computer at both ends, or a combination of both, man and machine.

From the relatively short descriptions that have been given for systems which utilize a space segment, it is evident that there are inherent basic similarities in the technologies associated with space exploration and space exploitation. These centre around the telemetry, tracking and telecommand requirements that provide the control facilities for satellite operation.

The design of the Earth segment should satisfy the requirements of the Space Segment, or vice versa, dependent upon inter-segment liaison requirements. The design of the space and Earth segments cannot produce an efficient system if undertaken independently. This means that if a literal top-down approach was adopted (space segment to Earth segment), the integration of a complete system may prove to be less than optimum.

1.5 SYSTEM PROGRAMME

For the example expounded in the book, it is assumed that initial designs have revealed that the utilization of a space segment would provide the best solution for a communication system.

The results of this initial system design are summarized in Fig. 1.4. This shows the programme that will need to be managed with skill to enable a communication system to be constructed and utilized. The figure also shows the timescales of a comprehensive programme that will produce an operational communication system of the type envisaged. The periods given are only an indication and depend upon how

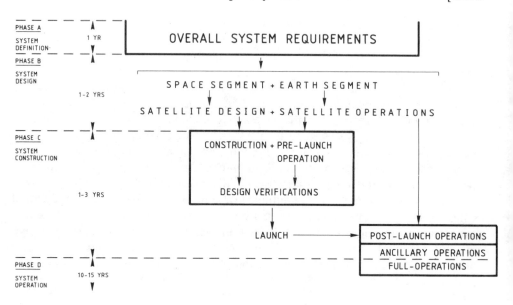

Fig. 1.4 — Complete overall communication system programme. Phases and timescales which require stringent management control to produce and operate an efficient system.

the requirements are defined and met. The programme illustrated consists of the following phases:

(a) Phases A and B. These are the system definition and design phases. Together they can take 1 to 3 years to complete.
(b) Phase C. This is the constructional phase of the system. It includes the launch followed by ancillary orbital operations and finally, commissioning of the satellite in its operational geostationary position. This phase can also take approximately 3 years to complete and includes any Earth segment construction and verification activities.
(c) Phase D. This is the operational phase which commences after the satellite has been commissioned in its operational orbit. The most likely terminating event for system operations will be the planned end of the satellite life. If all goes well, this will be due to the depletion of on board satellite fuel. The duration of this phase, if everything operates successfully, can be between 10 and 15 years.

The aerospace design should embrace the use of existing and standard elements, particularly concerning the Earth segment. New designs and developments will primarily be concentrated in the space segment. These may also necessitate changes to the Earth segment designs. However, before the satellite is launched, the overall design should be tested and verified to confirm that operational requirements can be achieved. Some elements within the satellite could be a re-use of previous develop-

ments which has been the case for programmes undertaken by the following
authorities:

INMARSAT	— ESA MARECS Spacecraft
EUTELSAT	— ESA ECS Spacecraft
FRENCH NATIONAL	— TELECOM 1 Spacecraft

Before the overall communication system becomes fully operational, the satellite
needs to be operated in an ancillary and preparatory manner immediately after its
launch. These satellite operations conclude with the commissioning of the complete
communication system as the penultimate to full operational status being achieved.

Operational activities for a spacecraft engaged upon an exploration mission do
often commence immediately after launch. In such a case, the space segment
constructional activities will terminate when the required flight path has been
attained. This can be achieved without any major spacecraft ancillary operations
such as commissioning activities being undertaken. Therefore, full operations can
commence immediately after the launch vehicle has placed the scientific spacecraft
into its flight path.

The reader will observe from Fig. 1.4 that a programme for the construction and
operation of an overall communication system of the type considered can run for a
period of 20 years. This time span is the life-cycle of the satellite, covering
requirement definitions and end-of-life in orbit. During most of this period, Earth
segment facilities will be required so that the satellite can be controlled and utilized.
Indeed, the overall system programme will exhibit many common facets which need
to be managed with expertise to enable efficient operations to be achieved.

Construction of the space segment can be undertaken with some independence
from the construction of the post-launch Earth segment. This being the case, the
space segment and the post-launch Earth segment should be verified together before
the space segment is launched. The operations methodology that has been employed
for verifying the post-launch Earth segment for the majority of European communication satellites is described in Chapter 3.

1.5.1 Operational organization

The basic architecture for a fully operational communication system employing a
spacecraft is given in Fig. 1.5. The figure depicts the post-launch configuration with
the spacecraft in its final orbit, which is a geostationary Earth orbit (GEO). This is
the fully operational system architecture for the subject communication system that
should meet the user requirements if it is operated correctly and efficiently. For
example, the users will communicate with each other over the RF links that are
provided by the geostationary satellite and associated Earth segment facilities to
enable telephone conversations to take place. The supplier of the service will test and
monitor the operation of these RF links with specialist support equipment that
operate under the control of terminal equipments commonly addressed as specialist
workstations. Workstations are control facilities that provide the manual interfaces
to the associated equipment. Similarly, the service supplier will ensure that the
satellite is controlled and configured efficiently to meet the user requirements. This is

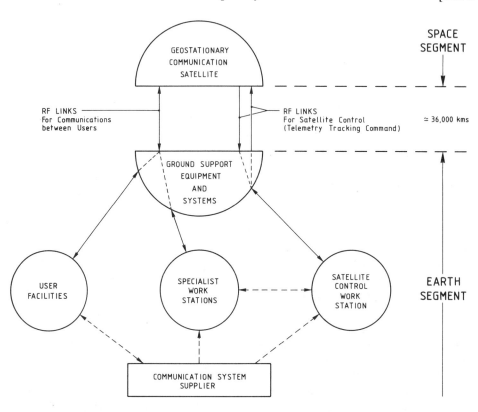

Fig. 1.5 — Basic architecture for an operational communication system. Infrastructure depicting Earth and space segment liaisons that enable user requirements to be achieved.

achieved with equipment that allows the satellite to be monitored and controlled by telemetry and telecommand signals. The Earth segment may also need to be reconfigured by the service supplier to meet user requirements.

Another facet of the control function will entail tracking and determining the orbit of the satellite. Ground support facilities will be required to perform these Telemetry, Tracking, and Command (TTC) functions. The support equipment associated with satellite control is operated at a satellite control workstation. The specialist and control workstations will be under the executive of the supplier of the communications system. There can be direct and indirect liaisons between the workstations during some operations. Also, users (customers) may give the communications supplier direct and indirect requests and requirements which will be relayed to either or both workstations. A satellite control workstation in such a communication system is usually assigned an executive control role. Consequently, the control workstation, together with the associated ground support equipment, is often termed the satellite Operations and Control Centre (OCC). It can and should delegate authority to other components of the Earth segment if efficient operations are to be performed.

The communication system which has been described is used as a prime example for demonstrating, in a fundamental manner, how systematic engineering techniques can be applied to spacecraft operations and control to give an efficient solution that can be commercially attractive. The communication system that employs satellites which were constructed and launched under the European Space Agency's (ESA) European Communication Satellite (ECS) programme are taken as the model.

1.6 SYSTEM ASPECTS CONSIDERED

The book begins by explaining the composition of an aerospace system and its component parts. This is followed by descriptions associated with spacecraft and their integration and testing. After this, spacecraft operations are addressed and compatibilities between system components utilized during pre- and post-launch phases are reviewed. The results of this review demonstrate the technical areas where rationalization can be established. In conclusion, evolving aerospace systems are discussed in the context of rationalization and cost effectiveness, resulting in the possibility for better engineered solutions that meet system requirements.

The book reviews the facilities and methods which have been adopted in Europe for the operation of unmanned spacecraft. The launch of a spacecraft is a short but crucial phase of a construction process which enables an aerospace system to become fully operational. In Europe, spacecraft operations have been:

- instigated during spacecraft construction,
- maintained up to and throughout launch activities,
- encompassed the full in-flight operational status,
- terminated at spacecraft end-of-life in orbit.

Thus, operational activities are practiced over many years.

The book demonstrates that multiple development activities deserve attention and can be reduced by a rationalized approach. The rationalization should result in standardized technologies and common operational practices. Such an approach can be regarded to be comprehensive and systematic if viewed over a complete spacecraft programme, commencing with technical definitions and embracing the full operational period. As part of the review process, the book proposes that spacecraft operational methodologies for pre- and post-launch operations can be harmonized, resulting in the re-use of technologies within and between programmes. Indeed, the equipment designs that have been produced to test and operate the first EUTELSAT (European Telecommunication Satellite Organization) satellites before launch have been employed on other programmes. The European Space Agency's ECS (European Communication Satellite) programme produced the first series of satellites for EUTELSAT. The equipment re-use has primarily been associated with pre-launch tasks, although they have been employed to a certain extent for in-orbit activities, as will be explained. The French National Telecommunication programme and INMARSAT (International Maritime Satellite Organization) have employed the same basic architectural components as ECS for the pre-launch operation of their spacecraft.

A communication system aligned to the services which are expected to be provided in the 1990s by satellites in geostationary Earth orbits are employed as a background medium for the reasoning of the book. Other types of spacecraft will exhibit a number of similarities with an aerospace system that meets a purely communication requirement. The European ECS programme is taken as an explanatory baseline by the authors. Satellites of this series were developed from one of the first European communication satellite programmes OTS (orbital test satellite) that was undertaken by the European Space Agency (ESA). Furthermore, some references to ESA OLYMPUS-1, one of the largest communication satellites ever launched, are also made. A comprehensive programme for such satellites can necessitate that they are operated and controlled for a 10 to 20 year period. Thus, a control system for a geostationary communication satellite is taken as the prime example. A satellite in a geostationary Earth orbit revolves at the same angular velocity as the Earth positioned at a nominally fixed equatorial longitude. Consequently, to observers on the Earth the satellite appears to be stationary and can be 'seen' from an arc of approximately forty percent of the equatorial circumference. The operation and control of a satellite in such a geostationary orbit can be undertaken from support equipments that are located at a single position on the Earth.

The book does contain some comparisons and analogies with other spacecraft types, and uses as an example a specific scientific spacecraft probe which was associated with Halley's Comet in the mid 1980s. Activities which were undertaken to meet this exploratory probe were performed as part of the ESA GIOTTO programme. Such comparative reviews should demonstrate that commonalities exist for space exploitive and exploratory missions.

The book highlights the similarities and differences between satellite operations before and after launch. Although the ultimate purposes of aerospace systems may differ, similarities are inherent, especially in the satellite operations field.

2

The space segment:
composition and construction

A space segment can be composed of a single spacecraft or several which operate together as part of an overall aerospace system. Before any system can be operated efficiently it must undergo a coordinated construction and test phase. Spacecraft can be tested and operated in a manner that simulates as closely as possible their expected launch and in-flight operational life. Therefore, spacecraft can be operated before and after their launch for space segment constructional purposes.

Support systems are required for all phases of space segment construction. The initial phase of this activity will be the building of a spacecraft. During this initial phase, support equipments are necessary for spacecraft testing and its preparation for launch. After it has been built and successfully verified, a launch vehicle will place the spacecraft onto its flight path. After launch, equipment can be required to make an overall aerospace system fully operational and to maintain it in this condition. This will be associated with the in-flight operations and control of the space segment. Sufficient coordination should allow any commonalities existing within the Earth segment configurations to be recognized and optimized in order to prevent any unnecessary duplications.

2.1 SYSTEM ASPECTS

The construction phase of the subject communication system can, as stated in Chapter 1, take up to 3 years to complete. The major space segment constructional activities are normally associated with the building of the communication satellite. This is followed by its launch and the early orbital activities which are a critical prelude to full system operations. The duration of the pre-launch constructional phase will depend upon how much design verification and testing is required for both the space and Earth segments. Should the Earth segment require extensive modifications then the most cost effective solution could be to modify the space segment design to cope with the Earth segment capabilities. There will be a requirement to verify that the post-launch Earth segment support facilities interface correctly with

the space segment. This must be performed before launch and at a time when the
spacecraft construction is in its final stages. The length and complexity of this phase
will be directly proportional to the degree of similarity in the design of the pre- and
post-launch support facilities.

A fundamental outcome of the definition and design phases of the subject
communication system programme has been that the space segment should be
composed of a single satellite in a geostationary earth orbit (GEO). Such an orbit will
allow the communication antennae on board the satellite to provide and maintain
communication links over fixed areas of the Earth. The terrestrial areas which can be
serviced by the beams of the communication antennae are often known as antenna
footprints. The antenna footprints for one of the first European communication
satellites OTS (orbital test satellite), which was successfully launched in 1978, are
shown in Fig. 2.1. This satellite model and the majority of European communication

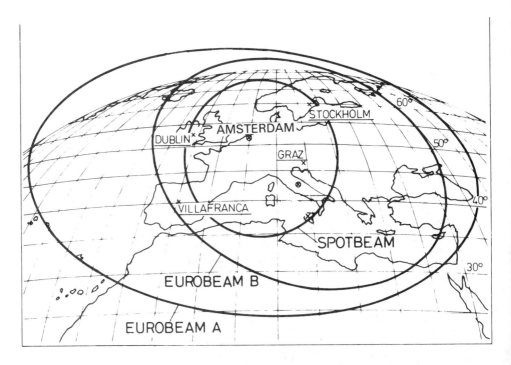

Fig. 2.1 — OTS antennae footprints. Geographical coverage of the first European communication spacecraft (courtesy of ESA).

satellites have provided communication facilities by operating in a body stabilized
(3-axis) mode. This means that the roll, pitch and yaw axes of the satellite are kept
constantly pointing in the same directions without any major movements, consequently the antennae is on a stabilized platform that remains almost stationary. This

is achieved with the facilities that are provided by on board satellite subsystems to provide the required antenna footprints.

2.2 POST-LAUNCH REQUIREMENTS

The configuration of the ECS series of satellites is shown in an exploded form in Fig. 2.2 which take ECS-1 as an example. ECS models are a direct derivative of the ESA OTS programme. A brief overview of the post-launch operations will now be given to enable the functional requirements of satellite subsystems to be more easily appreciated. Further explanations of these operations are given later in this chapter.

Before the fully operational status is achieved, there will be a requirement for some manoeuvring of the satellite in space. The first of these manoeuvres will be the separation of the satellite from the launch vehicle. After this has been achieved, the satellite should be travelling in an elliptical equatorial orbit. It is from these orbits that the satellite will need to be manoeuvred into its operational geostationary position. These orbits are generally termed transfer orbits or sometimes more specifically geosynchronous transfer orbits (GTO). The transition to its GEO will require the use of a rocket motor that is an integral part of the satellite. This motor, owing to operational requirements, is termed the apogee boost motor (ABM). After separation from the launch vehicle, the complete satellite should be spinning. In this situation the satellite is a rotating object, and tends to maintain a stable axis because of gyroscopic effects. The satellite is regarded as being in a simple spin stabilized mode of operation which is ancillary to the full operational condition. This method of stabilization will continue during its primary equatorial and elliptical orbits. At an apogee (greatest orbital distance from the Earth) of the GTO, the ABM will be fired so that the velocity of the satellite will be increased to boost it into a geosynchronous orbit. These orbital flight paths are shown in Fig. 2.3.

The transition shown in Fig. 2.3 portrays operations with the single burn of a solid fuel ABM. Liquid fuel ABMs can be fired several times, enabling the geostationary orbit to be attained in progressive stages. Such operations can enable this to be achieved as shown in Fig. 2.4, with the initial transfer orbit TR1 being achieved by the launch vehicle. The successive transfer orbits TR2 and TR3 are the results of ABM operations as is TR4, the transfer to the geostationary orbit. This second method of operation enables corrections to be made if unsatisfactory transfer orbits occur. After the geostationary orbit has been achieved, other satellite manoeuvres will be necessary so that the geostationary position can be attained.

When the geostationary orbital position has been acquired and 3-axis stabilization has occurred, the solar array wings, which are folded and mounted on the north and south pointing sides of the satellite, will be deployed. The attitude of the satellite will be controlled for antennae pointing purposes, but such actions, owing to satellite design will allow the solar arrays to be kept pointing along the pitch axis. After these manoeuvres have been completed, the satellite's communication subsystem is configured to establish the communication links.

The communications antennae are maintained in a stable Earth pointing mode. Fig. 2.2 shows that the antennae will be pointing earthwards along the +Z direction of the yaw axis when the communication system is fully operational.

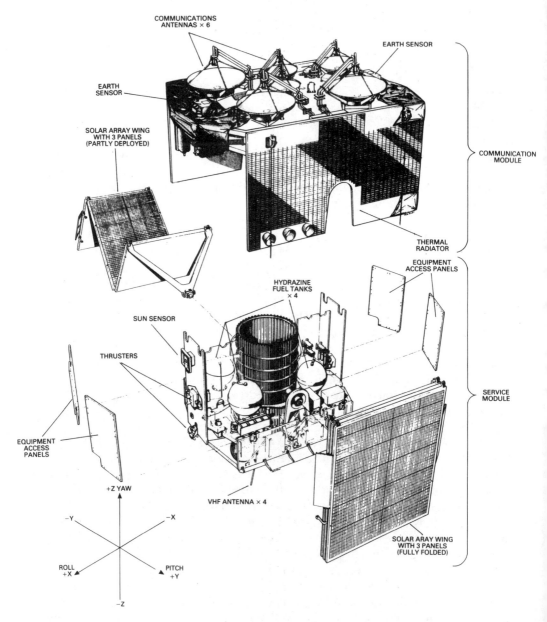

Fig. 2.2 — ECS-1 configuration. The architecture of all ECS models are identical in principle, but communication payload antennae differ (courtesy of ESA).

2.3   SATELLITE COMPOSITION

The satellite is composed of several subsystems which enable the fully operational status to be achieved and maintained. These ECS subsystems are now identified and described.

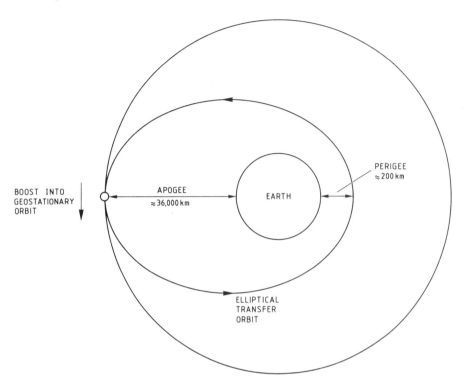

BOOST INTO
GEOSTATIONARY
ORBIT

APOGEE
≈ 36,000 km

EARTH

PERIGEE
≈ 200 km

ELLIPTICAL
TRANSFER
ORBIT

Fig. 2.3 — Geostationary orbit acquisition — solid fuel ABM. This is achieved with the single
firing of the apogee boost motor.

2.3.1   Telemetry, tracking, and command (TTC) subsystem

The TTC subsystem centralizes the encoding of information from all other on board
subsystems to enable these data to be transmitted as a telemetry signal. It is often
termed housekeeping telemetry data. The subsystem allows tracking functions to be
performed so that the exact orbital position of the satellite can be determined and
maintained. The TTC subsystem also enables telecommands to be received, decoded
and executed. This subsystem interfaces with ground support facilities of the Earth
segment so that satellite operations can be monitored and controlled.

2.3.2   Power supply subsystem (PSS)

This subsystem provides all the others with the electrical supplies required for their
operation. The PSS units and associated control electronics ensure that electrical
supplies are made to the other subsystems on a common supply line (50 V bus). The
subsystem is composed primarily of solar cells mounted upon two arrays which
charge two nickel cadmium batteries.

The solar arrays are folded for the launch and transfer orbits, and are deployed by
firing pyrotechnic devices after the satellite is 3-axis stabilized. The solar array wings
are then operated in a Sun tracking mode so that they can generate the maximum
electrical power.

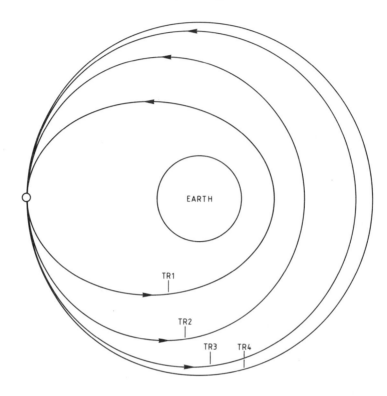

Fig. 2.4 — Geostationary orbit acquisition — liquid fuel ABM. This is achieved with several
successive firings of the apogee boost motor.

2.3.3   Pyrotechnic subsystem (PYRO)

The purpose of the pyrotechnic subsystem (PYRO) is to generate current pulses
following telecommand reception and route the pulses to the satellite pyrotechnic
devices. These devices allow the solar array wings to be deployed and the ABM to be
fired. The subsystem provides, via telecommands, arming circuitry so that devices
are isolated individually until immediately before the required operation time. This
means that two switches need to be closed to complete a circuit to enable a
pyrotechnic device to be fired.

2.3.4   Attitude and orbit control subsystem (AOCS)

The purpose of the AOCS is to provide the attitude determination and control
functions required by the satellite to achieve an operational geostationary orbital
position and enable the required antenna footprints to be acquired and maintained.
The subsystem must be able to perform functions in the transfer orbits to acquire the
geostationary orbit. To effectively perform these tasks, the subsystem must control
the spin-stabilized condition of the transfer orbits. After this condition has been
achieved and the ABM has been fired, the subsystem performs the transition to the
3-axis stabilized conditions required by the on-station orbital position.

The tasks outlined are performed by the subsystem operating gas jets with a monopropellant liquid hydrazine and nitrogen as the pressurant. The hydrazine and nitrogen (generally termed on board fuel) are loaded into on board satellite tanks. The liquid hydrazine flows under the pressure of the nitrogen after the necessary commanding action. This causes the hydrazine to come into contact with a catalyst, producing a chemical reaction that decomposes the hydrazine, forming a gas which is fired through jets which are termed thrusters. The firing of a thruster causes movement of the satellite. The thrusters are operated as a result of information obtained from sensors which respond to stimulation associated with the Earth, the Sun, and possibly the stars. Thus, the operation of sensors and thrusters enable the AOCS to perform the necessary actions that keep the satellite in a stable condition in its geostationary orbital position.

Should the satellite attain unwanted or dangerous attitudes, then the AOCS subsystem can take autonomous corrective action to place the satellite in a safe operational mode. This can result in antennae footprints being disturbed. However, telemetry and telecommand liaisons are maintained so that footprint recovery actions can be undertaken from the ECS ground support facilities.

2.3.5 Thermal control subsystem
This subsystem provides the operating thermal environment for all on board subsystems. The subsystem consists of mainly passive devices, including the thermal blankets which are fitted to the exterior of the satellite. The subsystem is also equipped with controllable heaters which are automatically operated by thermistors of the subsystem. These autonomous actions are undertaken to maintain safe operating temperatures of subsystem units. Some heaters are also operated under the control of the ECS ground support facilities via the telemetry and telecommand interfaces provided by the satellite TTC subsystem.

2.3.6 Communications payload
The communication subsystems on board the satellite are often termed the satellite payload (P/L). It is this subsystem which dominates satellite designs so that the fundamental objectives of users can be met by spaceborne antennae and repeaters. All other satellite subsystems support this communication subsystem so that links between Earth Segment elements can be achieved. Furthermore, when the satellite is operational, that is on-station, the RF interfaces for TTC subsystem links to the ground support facilities of the Earth segment can be provided by the payload for telemetry, tracking and telecommand purposes.

2.3.7 Apogee boost motor (ABM)
For the ECS programme, this was a solid fuel rocket motor which is operated once only after the satellite has been launched. The ABM enables the satellite's orbital path to be changed from equatorial elliptical (transfer) to geostationary (operational). This motor, although integrated into the satellite, is not generally considered as being a satellite subsystem because it is an integral part of the satellite for only a few days before launch. Furthermore, its operational function is maintained only for approximately one minute.

2.3.8 Satellite subsystem interfacing

The satellite subsystems will interface with each other to form the satellite system. The electrical interfaces are shown, in principle, as a block diagram in Fig. 2.5 which takes the ECS series as the example. It will be seen that the electrical power from the PSS is delivered on a 50 V bus. The health of each subsystem is monitored and controlled via telemetry and telecommand signals. The control of satellite health is maintained under the executive of the Earth segment via the TTC subsystem, which is required to operate according to inter-segment interface standards.

The subsystems that service the communications payload are the major constituents of the service module which is indicated on Fig. 2.2. Similarly, the communications module is comprised mainly of payload components. This architecture has enabled the service module to be constructed on a bus basis, enabling it to be re-used to support other payloads. The French national telecommunication satellite Telecom 1 is an example of an ECS service module design re-use as are the maritime communication satellites that were constructed under the auspices of ESA.

2.3.9 Spacecraft commonalities

The composition of a spacecraft engaged upon a scientific probing mission can have similarities with an application satellite. A scientific spacecraft will consist, in concept, of similar subsystems which, dependent upon its mission, may exhibit some common requirements with a geostationary communication satellite. The scientific experiments will need to be supported by the other subsystems which perform the housekeeping tasks. Therefore, some scientific spacecraft subsystems will be required to perform the same functions as those required by application satellite subsystems. Certainly, electrical power will be required, and solar cells and batteries can be configured to meet the demands for some scientific missions. The flight paths and orientation of the spacecraft will need to be controlled as in the case with geostationary applications satellites. Furthermore, telemetry and telecommands associated with housekeeping tasks will also exhibit some general similarities. In place of the communication payload, the scientific spacecraft will carry scientific instruments as its payload. These instruments are most generally termed experiments. The experiments may be scientific probing instruments which perform observations. The GIOTTO spacecraft that was engaged upon the probe into Halley's Comet in the mid 1980s had such types of experiments on board. Although the GIOTTO spacecraft was engaged upon a probing mission, similarities in spacecraft architectures can be identified from Figs 2.2.(communication) and 2.6 (scientific). For example, the kick motor (Fig. 2.6) was employed for flight path modifications, which in concept, is a similar function to the ABM of the ECS Series.

2.4 CRITICAL CONSTRUCTIONAL PHASES

The spacecraft launch and manoeuvring to its operational flight path followed by commissioning tests, are the critical and final constructional phases of the space segment. These activities are very much shorter in duration than the assembly, integration, and testing (AIT) of the spacecraft, which commences many months

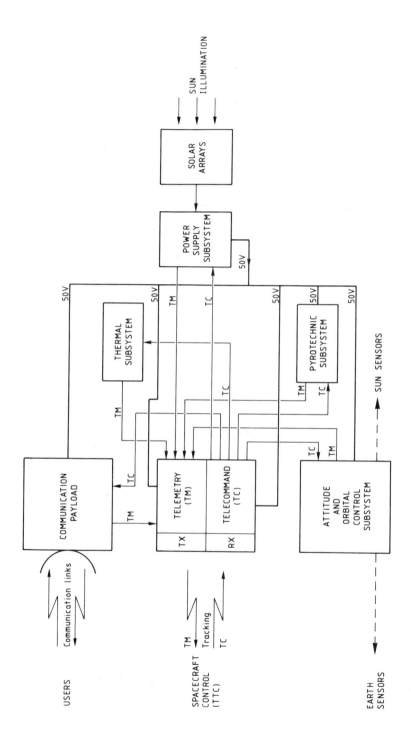

Fig. 2.5 — Satellite system block diagram — ECS example. The interconnections between satellite subsystems are shown, together with interfaces which are external to the satellite.

before a launch. However, if a small margin of error is exceeded during a launch, the results can be disastrous.

The launch operations of a spacecraft in the strictest definition commence when the launch vehicle engines are started. Launch operations terminate after the launch vehicle has placed the spacecraft onto a flight path that is free from the strong terrestrial gravitational forces. The flight path of the launch vehicle, that is from vehicle lift off until free from the major influences of the Earth's gravity, is commonly termed the launch trajectory.

2.4.1 Launch activities

The duration of an expendable launch vehicle (ELV) trajectory for placing a spacecraft into an orbit from which a transfer to a geostationary position may be achieved is in the order of 30 minutes. Thus, the ELV plays a prime operating role in the construction process of the communication system for a short period.

From a more general aspect, satellite launch operations commence earlier than ELV engine start up. The satellite has to be prepared, mounted, and made ready for its flight upon the ELV. The final phases of satellite activities are associated with loading on board fuels and integrating the ABM with the satellite. After these tasks have been performed, the satellite is mounted upon the launch vehicle. For all these final activities, the executive control of the satellite is maintained by the electrical ground support equipment (EGSE) utilized during its construction. It is this EGSE that gives a GO/NOGO for launch, dependent upon the satellite status.

The launch activities place the communication satellite in operational modes which are preparatory to its fully operational status. These modes must be successfully accomplished, if a fully operational space segment is to be realized.

There is ground support equipment that services the ELV during its construction, launch, preparation, and flight. In an active operational role, the equipment provides a monitoring service for the ELV during its flight which, if it is unsatisfactory, can result in the ordering of its destruction. Such an order to cause ELV destruction is given if the launch trajectory is not as expected and can result in dangerous consequences if allowed to continue. The danger could be that debris returns to Earth in a position that could endanger human life where it may fall.

It is quite common for the operation of an ELV to be in two or three stages, dependent on its architecture. A two stage ELV is composed of two prime separate parts, and, obviously, there are three prime separate parts for a three stage ELV. Fig. 2.7 shows the architecture of the European three stage ELV Ariane-4, configured for a dual spacecraft launch. The first stage of this ELV model is equipped with additional engines which are termed boosters. These give assistance to the first stage engines for the initial thrust which is required for lift-off at launch.

To overcome terrestrial gravitational forces the first stage engines are the most powerful. When they and their boosters have lifted the ELV and its satellite cargo to an altitude of approximately 60 kilometres in about 2 minutes, the first stage separates from the other two stages. The second stage engines then fire and at the end of their operation, the second stage separates from the third stage. When this occurs, the third stage engines fire and operate until the transfer orbit is achieved whence the satellite is ejected from this final stage of the rocket. If all has operated correctly, the satellite will be placed upon the required flight path. Table 2.1 presents an overview

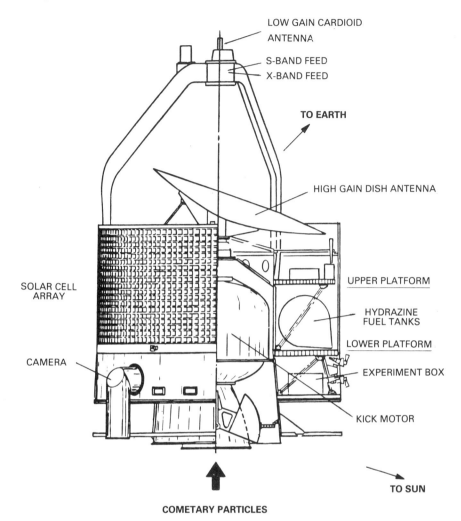

LOW GAIN CARDIOID ANTENNA

S-BAND FEED
X-BAND FEED

TO EARTH

HIGH GAIN DISH ANTENNA

SOLAR CELL ARRAY

UPPER PLATFORM

HYDRAZINE FUEL TANKS

LOWER PLATFORM

CAMERA

EXPERIMENT BOX

KICK MOTOR

TO SUN

COMETARY PARTICLES

Fig. 2.6 — Architecture of the ESA GIOTTO spacecraft. This spacecraft which made a scientific probe into Halley's Comet had commonalities with a geostationary communication satellite (courtesy of ESA).

of the basic schedule of activities for correct operation of Ariane-4 after the ignition of the first stage engines and boosters.

If each stage of the launch vehicle operates correctly, the engines will cease to function when fuel is expended. When the first stage engines burn-out it will separate from the second and third stages. After its separation it will free fall in space and burn up by friction within the Earth's atmosphere. For the second and third stages this may not be the case; then the stage will orbit the Earth before its flight path decays sufficiently for it to enter the atmosphere, and then the burn-up process commences. This is what normally happens with the third stage. Thus, because the stages are allowed to be expended, a rocket is more commonly termed an expendable launch vehicle (ELV).

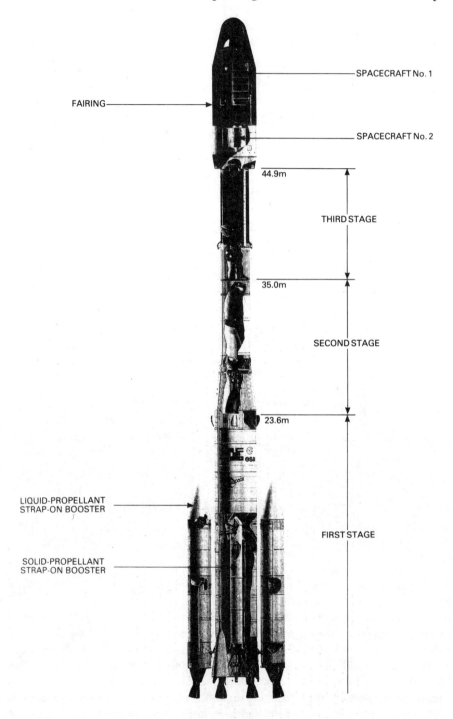

Fig. 2.7 — Basic architecture of Ariane. Configured for a launch of two communication
spacecraft; Superbird A for Japan and DFS1 for Germany (courtesy of Arianespace).

Table 2.1 — Basic schedule of activities for an Ariane 4 launch associated with a payload of two spacecraft

TIME	ACTIVITY
Ho	**First stage and booster engines ignition**
+ 3.4s	**Lift Off**
+ 15s	End of vertical ascent phase and start of pitch motion (10s)
+ 2 min 30s	Boosters jettisoning (1st pair)
+ 2 min 31s	Boosters jettisoning (2nd pair)
+ 3 min 32s	First stage separation
+ 3 min 34s	Second stage engine ignition
+ 4 min 21s	Fairing jettisoning
+ 5 min 44s	Second stage separation
+ 5 min 48s	Third stage engine ignition
+17 min 51s	Third stage shutdown sequence
+17 min 53s	GTO injection
+20 min 23s	**First spacecraft separation**
+25 min 05s	**Second spacecraft separation**
+25 min 10s	Start of the third stage avoidance manoeuvre
+29 min 11s	**End of ARIANE FLIGHT**

2.4.2 Satellite orbital phases

The final tasks associated with the construction of the space segment commences after the satellite has been successfully launched and ejected from the ELV. The European communication satellites (ECS) have been launched on Ariane and placed into near equatorial elliptical orbits after ejection from the launch vehicle. At this juncture, there are manoeuvres to be made so that the operational geostationary orbit can be acquired. For these manoeuvres the space and Earth segments will be in ancillary working modes from the overall communication system standpoint.

The operational working status of the communication system is when the satellite is in a geostationary orbit with its communication payload operational. The different modes of operation which the satellite is subjected to for a relatively short period after its launch, although preparatory and ancillary to the fully operational conditions, are vital to the achievement of the fully operational status.

The final tasks of space segment construction will be the acquisition of the fully operational orbit and satellite attitude that satisfies the communication system requirement, followed by satellite commissioning activities. These final tasks for the construction phase will be of relatively short duration but decisive in producing an operable system. They will impact sharply upon satellite designs and, therefore, from a top-down system engineering standpoint, these tasks will need serious consideration during the space segment and Earth segment design phases.

2.4.3 Launch and early orbit phase

A pictorial summary of the operational activities which are performed during the final phases of space segment construction are shown in Fig. 2.8 for an ECS communication system.

Satellite in-orbit operations actively commence with the acquisition of the telemetry signal (AOS) by a TTC station after separation from the launch vehicle. Dependent upon the location of the satellite TTC stations, AOS can occur between 20 and 30 minutes after lift off. The operations then continue for the ancillary activities during the early orbits. The initial phase of satellite operations is known as the launch and early orbits phase, the LEOP. The LEOP can be a period of several days, which includes all operations until the correct longitudinal position in geostationary orbit has been achieved. Afterwards, when the correct orbital position has been attained, commonly expressed as station acquisition, the satellite's payload is commissioned and the communication system can then become fully operational.

2.4.4 Transfer orbits

The transfer orbits will be elliptical and near equatorial with an apogee of approximately 36 000 kilometres and a perigee of approximately 200 kilometres. During these orbits, the satellite is not operational from a communication system standpoint. However, correct performance during its transition from the elliptical to the geostationary orbit is vital if it is to transfer to full operational status.

Ideally, as indicated in Fig. 2.3, the apogee of the transfer orbit should be at the same distance from Earth and at the same longitude as that required by the geostationary position. Thus, to achieve the required geostationary orbit, the satellite should remain at transfer orbit apogee. This could be achieved by the single firing of solid fuel ABM to circularize the orbit and achieve a geostationary position. After the operation of the ABM, the satellite should then remain at a fixed position above the Earth. However, the elliptical transfer orbit that is achieved is greatly dependent upon the performance of the launch vehicle and the location of the launch site. Because of small perturbations in the orbital path, it is normally necessary to trim the satellite's orbital geometrics under ground control before firing the ABM. During its transfer orbits, the satellite spin rate is increased by its AOCS thruster gas jets to approximately 60 rev/min to provide adequate gyroscopic stability for ABM firing. The firing of the ABM, which imparts a very high thrust, is performed at the transfer orbit apogee, reasonably close to the nominal longitude allocated for the geostationary orbital position. The satellite drifts in longitude with respect to the Earth at a rate of a few degrees each day. This period could be shortened, but at the expense of the on board AOCS fuel. The satellite will most likely drift for several days toward its required geostationary position before the exact geostationary orbit is acquired.

The satellite must be manoeuvred to its operational position with the use of the satellite's AOCS. Before performing final orbit corrections, it is necessary to configure the attitude stabilization of the satellite to its 3-axis pointing mode. This reconfiguration is a critical series of manoeuvres performed partly under control of the support facilities of the Earth segment and partly using the automatic logic of the on board satellite AOCS. The satellite spinning is stopped and the solar arrays, which had previously been stowed flat along the sides of the satellite for compactness

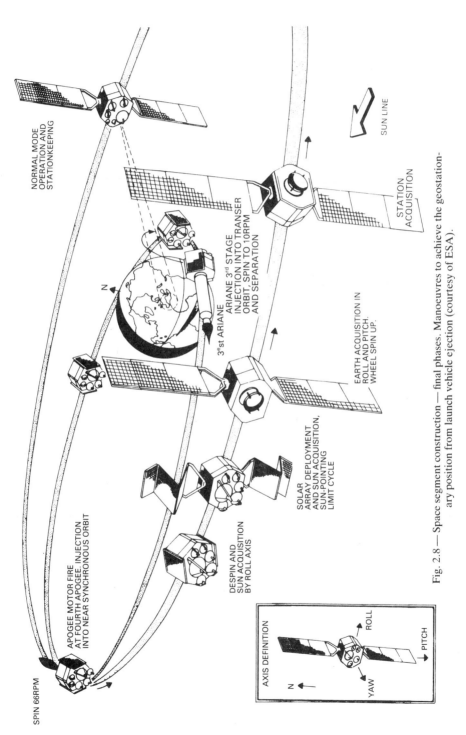

Fig. 2.8 — Space segment construction — final phases. Manoeuvres to achieve the geostationary position from launch vehicle ejection (courtesy of ESA).

during launch, are then deployed. This action is achieved by firing pyrotechnic devices which release springs that enable solar array deployment. The satellite is then manoeuvred so that the antennae are pointed toward the required part of the Earth's surface. The satellite is then stabilized in this position. If all these activities are successful, then the satellite will be positioned in a geostationary orbit at the required longitudinal station and will exhibit the required attitude to the Earth. This will allow the overall communication system to be commissioned. In its final 3-axis configuration, the satellite moves in a geostationary orbit with the communication antennae pointing toward the Earth. The solar arrays with correct operation of on board equipments face the Sun continuously to provide electrical power.

In summary, the transfer orbits are normally deemed to commence with the separation of the satellite from the launch vehicle and terminate when the operational orbit has been acquired (station acquisition).

Throughout all the activities performed during the transfer orbits, the health of the satellite is monitored and controlled by the ground support facilities of the Earth segment.

2.4.5   Geostationary orbit

After the successful completion of operations during transfer orbits, the satellite will be geostationary. A satellite when in a geostationary Earth orbit remains at a fixed equatorial position above the Earth at a distance of approximately 36 000 kilometres.

When the exact on-station orbital position has been acquired, the communication payload is switched on and commissioned. After this activity has been successfully completed, the space segment construction activities terminate. Services that are provided by the communication system can then be utilized. Changes in the configuration of the satellite communication payload may be necessary to satisfy user (customer) requirements. Furthermore, operational activities will be necessary from time to time to maintain the orbit and retain the antennae footprints over the required areas of the Earth. When the satellite is on-station in its geostationary orbital position and fully operational, the satellite operations can be predictable and routine if all is functioning correctly.

2.5   MAJOR CONSTRUCTIONAL PHASE

Pre-launch activities of space segment construction can extend over a significant period (1 to 3 years), hence the term major constructional phase.

The fundamental level of spacecraft construction commences with the building of equipments. This is generally known as unit level construction. The unit level equipments are first tested independently, then assembled into subsystems which are subjected to further testing. After this, the subsystems are mounted upon the mechanical structure to form the system which will be launched, the spacecraft.

The spacecraft structure can be considered as a subsystem and may have more than one component. As shown in Fig. 2.2, the ECS satellite structure is composed of two prime elements: the service module (SM) and the communications module (CM). The two structural modules are constructed independently and then integrated to form the satellite. When the satellite is complete, it undergoes a test programme that verifies that the on board satellite subsystems can operate together without any incompatibilities. The programme then goes on to simulate the expected environments which the satellite will experience during its launch and in-flight life. This is a qualification process to prove that the spacecraft is ready for its launch and working life.

During the qualification process, the space environments need to be simulated by test facilities and chambers. Similarly, electrical test equipments are necessary to complement these environmental facilities during the testing of the spacecraft. Spacecraft control is performed by telemetry and telecommand transmissions when the spacecraft is in-flight. Because of this, exactly the same methods are utilized during construction and test.

A significant number of construction activities take place in a laboratory environment. The laboratory environment is strictly controlled regarding temperature, humidity, and cleanliness. This is because some units of the spacecraft can be delicate, and if environmental limits are exceeded, damage can occur. As mentioned previously, the satellite testing activities will include environmental tests that simulate launch and post-launch operations. A fundamental difference between pre- and post-launch activities is the distance between the satellite and the pre- and post-launch Earth segment elements. This will necessitate differences in RF power requirements for the telemetry and telecommand communications between the space and Earth segments of an aerospace system.

For the subject communication satellite, post-launch operations can be envisaged to take place over a period of at least a decade if all operates successfully. Pre-launch operations of the satellite should include rehearsals of post-launch activities and include worst case situations. If all goes well when the satellite is in orbit, there will be the possibility for less satellite control activities than during its testing and pre-launch verification. Some aspects of the post-launch activities will not be able to be exercised to the full extent during satellite construction. Orbital determinations by tracking techniques are a case in point. Nevertheless, the on board satellite equipment which is utilized to perform the tracking activity will be functionally tested and verified.

The task of integrating and testing spacecraft subsystems to form the spacecraft is often known as spacecraft level (system) assembly, integration, and testing (AIT). In some cases, a complete functional test is not necessary during spacecraft system level AIT. A verification of operation can be sufficient if a full performance test has been made during unit or subsystem level testing. In such cases, the spacecraft system level operation can be termed a verification, hence the emergence of the term spacecraft level (system) assembly, integration, and verification (AIV). AIV can be considered to be a derivative of unit or subsystem level AIT.

2.5.1 Historical testing and verification philosophy

In the 1970s it was not uncommon for several spacecraft models to be built as developments for the flight model that would be launched. These development models could be separated into structural, thermal and electrical models. From a spacecraft operation standpoint, the electrical models which were operated by telemetry and telecommands could be termed:

Engineering model (EM)

Constructed with components that would not be subjected to expected launch and space environments. Most activities with EM spacecraft were conducted in a laboratory environment. This model was constructed to enable electrical designs to be verified at system level. Any design anomalies would be corrected and implemented on the next development model, the qualification model.

Qualification model (QM)
Constructed with components that would be subjected to environmental conditions that are greater than those expected. These conditions were relevant to launch and the non-terrestrial environments. QM operations were a stage of total system level design qualification which included thermal and structural aspects.

Flight model (FM)
Constructed and verified at lower environments than the QM so that spacecraft would not be overstressed before experiencing the full working situation. The flight model is the operational spacecraft which would be launched.

2.5.2 Prototype flight model philosophy
With the experience gained from historical testing methods, it is now possible to construct prototype flight model (PFM) spacecraft. As the name implies, these are prototypes which will be launched, and therefore, obviate the need for the construction of EM and QM spacecraft, thus giving cost saving incentives. A PFM approach to spacecraft construction and pre-launch operations can, if efficiently implemented, facilitate cost-effective production of an overall system.

The PFM spacecraft undergoes the same basic tests as those of the EM and QM, but perhaps not to the same upper levels as the QM satellites. When the satellite is complete, these tests are a verification process, that is, testing is at a level considered appropriate as a qualification to prove readiness for launch. A PFM spacecraft will first be operated in a laboratory environment and then, if this is successful, a repeat of the operations will be made in the differing and more demanding environments of launch and space simulations.

A PFM approach has been undertaken for the majority of European communication satellites, such as ESA's OLYMPUS-1 launched in mid 1989.

These satellites have been operated in simulated launch and space environments which have normally included two prime environmental tests.

Vibration tests
These tests simulate the expected vibrations that will be experienced by the satellite during launch. Vibrations caused by the operation of the launch vehicle engines can be transmitted to the satellite mechanically and acoustically.

The satellite is attached to the necessary vibration facilities in a manner that is representative of its mounting on the launch vehicle, and is subjected to vibration levels that are representative of those induced by the launch vehicle. For these tests, the satellite on board subsystems will be in launch configuration.

Thermal vacuum tests
The environmental conditions of a vacuum with solar simulations are produced in a test chamber. The satellite is placed in the chamber which is pumped down to a vacuum. Lamps provide illumination which is a simulation of the Sun. For some of the tests, the solar simulators are not used, thereby simulating a shadow condition for

when the satellite is not exposed to the Sun. Such events occur during transfer orbits or when the satellite is subjected to an eclipse of the Sun after the geostationary orbital position has been achieved. The environmental variations follow an operational sequence that is representative of the expected conditions that will be experienced during transfer and geostationary orbits. The simulations are not normally exact from all standpoints, for example the solar arrays are not deployed. Furthermore, the periods allocated to the different orbital conditions are shorter than the predicted post-launch expectations. On-board satellite redundancies and failure recovery modes are normally exercised, and if all goes well after launch, some of these exercises may not be performed during orbital operations. In summary, thermal vacuum tests necessitate placing the satellite in a vacuum and varying the temperatures over a range, typically 80°C to −130°C for geostationary orbits.

PFM testing philosophy can be equally applicable to a spacecraft that will be launched upon a scientific probing mission into space. However, in such a case, all in-flight activities will not be exercised because exploratory probing missions can be venturing into the unknown, making complete simulations of environments which the spacecraft will experience impossible.

Throughout all phases of the space segment construction, support facilities are necessary. This means that space segment construction activities require support facilities before, during, and after launch. These facilities have always been of prime importance for aerospace system safety and efficiency. Therefore, the Earth segments, which include the support equipment, should be given consideration during the design activities of an aerospace system. This should allow rationalization to be achieved, enabling equipment standardization and re-use during the construction and operational phases of a complete aerospace programme.

3

The Earth segment:
space segment support

The space segment of an aerospace system needs to be supported for both the pre- and post-launch activities. There will be commonalities and even an amount of exact duplication of requirements for pre- and post-launch Earth segments. These commonalities are centred around the control aspects of spacecraft operations. The communication system that embraces a geostationary satellite as the space segment is used to explain the principles of satellite operations; however, these principles can be applied to other spacecraft types.

Before proceeding further, a review and some definitions need to be made regarding what constitutes a post-launch Earth segment. This term is an expansive definition encompassing the facilities that are required for spacecraft control and those required by spacecraft users. For the post-launch activities, the major facilities required for spacecraft control are those that provide the capability for telemetry, tracking, and command (TTC). These spacecraft support facilities are provided by Earth segment elements that can be termed TTC stations. The TTC stations are operated under the centralized control of an operations and control centre (OCC). These basic support facilities are generally known as the ground segment, and this is the definition which will now be employed predominantly in this book. The other major elements of the complete post-launch Earth segment are the user facilities. For the subject communication system these user facilities are often referred to as Earth stations. Fig. 3.1 is an overview of this type of configuration.

These definitions are also applicable, in principle, to a scientific spacecraft operational scenario. The ground segment provides the TTC capabilities under the executive control of the OCC. The user facilities can be intelligent workstations that incorporate microprocessors which can also be interfaced to the OCC. Such a set-up was made, in principle, for the GIOTTO spacecraft operations, and the liaisons with the OCC enabled telecommand requests to be actioned and telemetry data to be supplied to the user facilities. The GIOTTO Earth segment comprised user workstations and a ground segment. An overview of this Earth segment architecture is shown in Fig. 3.2.

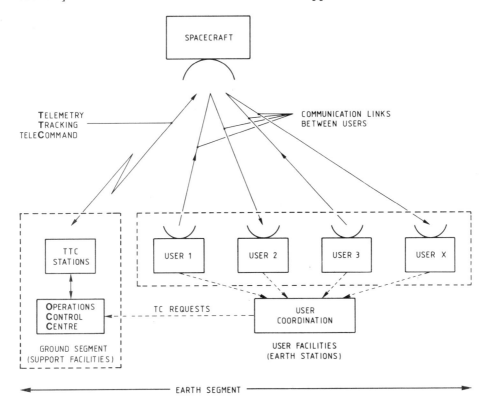

Fig. 3.1 — Earth segment for communication satellite. The support facilities enable the user requirement to be met when the geostationary position has been achieved.

3.1 COMMUNICATION SATELLITES SUPPORT

An Earth segment for the communication system will be required to support the spacecraft operating in various environments before it is operational as an on-station geostationary orbiting satellite. These environments necessitate that the satellite has several modes of operation. Working from the fully operational status, with a top-down systematic approach, these differing operating modes and environments will be driven by design requirements that should evolve from the:

- operational geostationary position,
- geostationary transfer orbits,
- satellite launch, and
- satellite construction.

To cope with these activities, support facilities will be necessary. These facilities will be configured to cope with activities that commence during satellite construction and terminate at the end of the satellite's operational life.

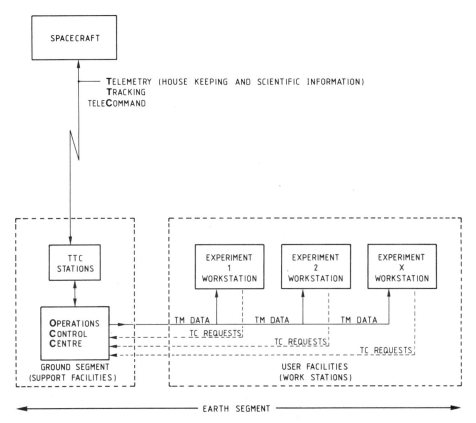

Fig. 3.2 — Earth segment for scientific satellite (GIOTTO). The support facilities operated in conjunction with the experiment workstations so that the scientific probe could be efficiently operated.

If all goes well, the satellite's operational life can last for at least a decade, and even beyond its expected lifespan. Throughout this period, its status is monitored and controlled so that user requirements can be met and maintained.

To perform all the necessary activities associated with post-launch operations, a space segment requires the support facilities of a ground segment. Similarly, for pre-launch activities, a space segment needs to be operated and controlled.

3.2 MISSION CONTROL ACTIVITIES

All the activities which are necessary to achieve and maintain a fully operational spacecraft are commonly termed mission control. Tasks will include orbit predictions and determinations. The mission control activities for an on-station geostationary communication satellite can be of a rather routine nature if all operates satisfactorily. The same category of activities will be undertaken during launch and early orbits phases, but during this period activities will be intense. In addition, it will be appreciated that mission control activities do commence before launch with orbit prediction tasks. It can also be necessary to perform tests associated with satellite control before launch to verify that the post-launch Earth segment is satisfactory to perform these tasks after launch.

Post-launch mission control includes configuring and monitoring the condition of the satellite. Other actions will enable the orbit of the satellite to be monitored and controlled by the orbit tracking activities. For the subject communication satellite regular routine tests will be performed upon the payload to determine performance criteria after commissioning activities have ceased. These tests can be undertaken with the active support of a single TTC station when the satellite is fully operational. The tasks which are part of the on-station mission control activities consist of the following:

- satellite health control
- flight dynamics
- trend analysis.

3.2.1 Satellite health control

For all post-launch operations, satellite health is the term for a satellite's correct and expected operation in the environment of space. The satellite's health is determined by the analysis of digital status parameters and analogue values which are transmitted by the satellite as telemetry signals. Control is achieved by telecommands to the satellite which can change its status and instigate an alteration to the orbital position. These results will be verified by telemetry analysis. Satellite control by telemetry and telecommand is essentially a closed-loop operation.

The monitoring and control of the satellite's health via telemetry and telecommand is normally carried out continuously for the duration of its life. If a telecommand is transmitted to the satellite, the responses in the form of telemetry data can be monitored and analyzed. By this means, the health of the satellite can be monitored as events occur and are experienced, thus allowing any necessary actions to be implemented promptly, that is to say in real-time.

3.2.2 Flight dynamics

The orbit determination, control, and verification activities are part of the flight dynamics task. Flight dynamics activities commence before a satellite is launched in order to predict the post-launch flight paths and orbits. Post-launch flight dynamics activities, which include orbital determination tasks, are complementary to pre-launch orbital predictions.

During the post-launch phase, the orbit of the satellite is determined by means of TTC station tracking facilities. A series of measurements are made by transmitting signals from these ground support facilities to the satellite and comparing the signals returned from the satellite. The time differences between transmissions from and receptions at the TTC station are the measurements which are made. These time measurements are processed by computer to determine the orbital position of the satellite. If the determination process shows that the satellite is not in a satisfactory orbital position then telecommands are sent to the satellite. The telecommands activate units of the attitude and orbital control system (AOCS) to make the necessary corrections to the orbit. Such actions are monitored by means of satellite telemetry in conjunction with continued tracking operations.

The transmission and reception activities of the tracking tasks are performed in real-time. The computerized orbital determination calculations, that is the analysis of the on-line tracking results, can be performed off-line at a later time by recording the results. Such a course of action enables the spacecraft orbit to settle before tracking activities recommence to verify that the corrective actions have been performed satisfactorily.

Flight dynamics activities include satellite attitude control tasks. These comprise monitoring and controlling the pointing of the satellite's three axes and, consequently, the position of antenna footprints.

Both the attitude and the orbit of a satellite are influenced by its dynamics. For example, the levels (and hence the weight) of on board fuel will decrease with in-orbit operations, thus affecting the dynamics of the satellite. This necessitates that the on board fuel levels be monitored and recorded so that calculations associated with the flight dynamics tasks can be meaningful. Such changes in satellite dynamics can have influences upon its attitude and orbit. Other factors influencing the dynamics and orbit of the satellite are the gravitational environment to which the satellite is exposed, and other conditions that exist in space. Therefore, flight dynamics tasks must continue for the full in-flight operational life of the satellite.

3.2.3   Trend analysis

During all post-launch activities, the satellite telemetry data which are received by the ground support facilities of the Earth segment will be archived on some form of storage medium such as magnetic tape or optical discs. These archives will form recorded history of the space segment's performance, and will cover many months and years of data collections. The long term archives can be processed to establish trends in satellite performance. This process is what is generally termed trend analysis. The results of this process can demonstrate the effects of ageing and the space environment upon the satellite. These results should be taken into account during the design and development phases of evolving aerospace systems. The results can also be of assistance for operational activities aligned to satellite health control. The ageing of a satellite unit can show a decline in performance which is detected by trend analysis. This can predict the necessity for the use of on board redundancy. Furthermore, trends in the use of on board fuel can be established and aid the flight dynamics operations. An example of the trend analysis associated with the on board AOCS hydrazine fuel diminution affecting satellite mass is shown in Fig. 3.3.

Usually, the activities which have taken place before the satellite launch are also archived by the pre-launch Earth segment. Access to this information can be of assistance to mission control activities. Such archives may not be part of the commonly practised trend analysis process which commences at the acquisition of telemetry signal (AOS) after launch.

3.3   POST-LAUNCH EARTH SEGMENT

The initial phases of post-launch activities will be associated with a complex of support facilities to cope with the operational requirements of the communication satellite. During the early phases of satellite operation, the Earth segment is

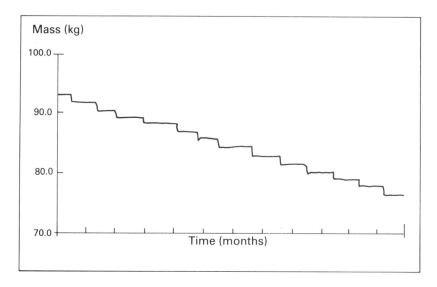

Fig. 3.3 — ECS-2 hydrazine mass reduction during a 12 month period. The weight of the AOCS fuel which is stored in tanks needs to be known so that flight dynamics tasks can be effectively performed (courtesy of ESA).

incomplete because the communication payloads are not operational, and therefore, user Earth stations are not employed. However, mission control activities during this period will be at their busiest, owing to the in-orbit manoeuvres which are necessary to place the satellite on-station. Once on-station acquisition has been achieved, mission control activities will be less intensive, and the Earth segment will be complete as user Earth stations will become operational when the communication payload is commissioned.

3.3.1   Ground segment

The support facilities which undertake post-launch control activities consist of components that are generally known as spacecraft ground stations. These stations, which include the executive controller of the space segment, comprise the following prime elements, termed the ground segment in the ESA telemetry and telecommand standards:

- Operations and control centre (OCC). This centre provides workstation facilities for satellite control. It has executive control over the other stations which are employed for spacecraft operations.
- Telemetry, tracking, and command (TTC) stations. They provide telemetry, tracking, and command interfaces between the Earth and the satellite. These operate under the executive control of the OCC.

During transfer orbits, manoeuvres will be made to attain the exact geostationary orbital position of the satellite. The satellite in transfer orbits, shown pictorially in

Fig. 3.4, takes a European communication satellite (ECS) as an example. The figure shows that the ground segment incorporates several TTC stations to monitor and control the satellite before geostationary orbit acquisition under the executive control of an OCC. The need for several TTC stations for these tasks is so that the ground segment contact with the satellite can be almost continuous throughout its elliptical equatorial orbits in order to track and control the flight paths. The tracking function is described in a little more detail in Chapter 4.

All TTC functions are undertaken at VHF (30–300 MHz) during transfer orbits. When the satellite is on-station and fully operational, these functions can be accomplished by using a channel of the communication payload.

3.3.2 On-station configuration

A geostationary position allows a single TTC station to provide interfaces to the satellite for control and monitoring purposes. When in its final geostationary position, the satellite should be able to meet the requirements of its users, but to do this, the satellite must be able to cope with its environment. There will be extreme differences in temperatures; the side of the satellite which faces the Sun will be very hot, whereas the side in shadow will be very cold. Additionally, the effects of the gravitational forces and solar winds will have impacts on the satellite's orbital position. Thus, from these considerations it will be appreciated why the satellite needs to be monitored and controlled during its orbital life. From an overall standpoint, the activities include configuring the satellite to meet users' requirements upon notification from the communication service suppliers. Efficient monitoring and control of the satellite enables the communication system to be fully operational. The final stage in the construction of the subject communication system is the testing and commissioning of the communication satellite in its operational geostationary orbital position. Terminals which are known as test and monitor stations (TMS) are utilized for in-orbit testing (IOT) of the communication payloads. The stations are used initially for commissioning tasks, and then to effect tests from time-to-time to monitor the performance of the payloads. The TMS operate under the executive control of the OCC for all major satellite IOTs.

The configuration of the complete Earth segment when the communication system is operational is shown in Fig. 3.5. A single TTC station controls satellite post-launch operations.

3.4 PRE-LAUNCH ACTIVITIES

Pre-launch activities of a space segment should include a verification of spacecraft design and operations under the simulated conditions of expected post-launch environments. Pre-launch operations of a fully integrated spacecraft can be summarized as being driven by post-launch requirements.

Initial spacecraft constructional activities can subject units to tests in a simulated space environment. After unit level tests have been undertaken, the units are assembled into subsystems and undergo a further test programme. Finally, the subsystems are assembled to form the spacecraft which is then subjected to tests that should verify its readiness for launch and operational life.

The following tests are undertaken as part of the pre-launch activities:

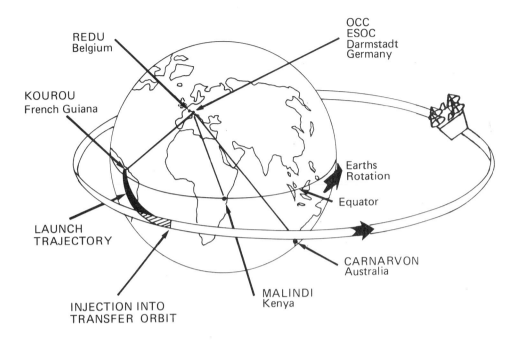

REDU
Belgium

OCC
ESOC
Darmstadt
Germany

KOUROU
French Guiana

Earths
Rotation

Equator

LAUNCH
TRAJECTORY

CARNARVON
Australia

INJECTION INTO
TRANSFER ORBIT

MALINDI
Kenya

Fig. 3.4 — Pictorial overview of telemetry data reception during transfer orbits — ECS example. An overview that shows the location of the ground segment elements for this phase of satellite operations (courtesy of ESA).

- Integrated systems tests (IST)
- Environmental tests
- Special performance tests.

3.4.1   Integrated systems test (IST)
This test is a verification of satellite operation for:

- launch mode,
- early orbits,
- geostationary orbit.

    The IST is conducted on a real-time basis and includes full on board redundancy tests and failure recovery modes.

3.4.2   Environmental tests
Two tests are carried out to represent a simulation of the expected launch and post-launch environments. These tests, described in Chapter 2, are:

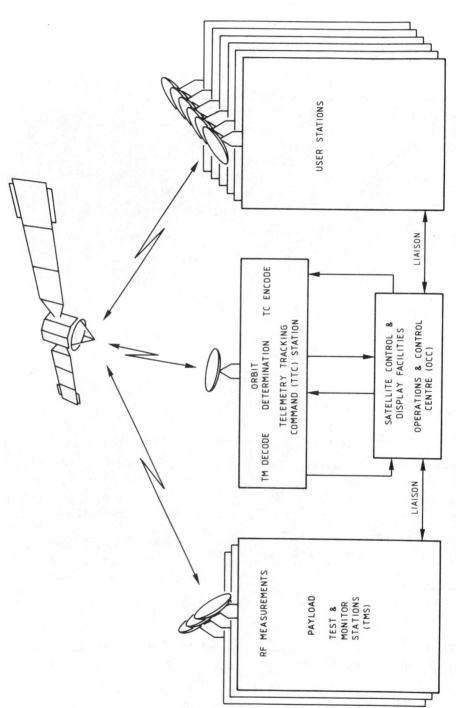

Fig. 3.5 — Earth segment for post-launch operations. The inter-segment liaisons for satellite on-station operations employ the ground segment to meet user requirements.

- vibration test to simulate launch,
- thermal vacuum test to simulate orbital life.

During these activities, IST operations are performed.

3.4.3 Special performance tests

These tests are verifications, and are complementary to the environmental tests and the IST. Some activities may need to be performed that enable on board unit level and subsystem level performance measurements to be made. Because all post-launch situations cannot be simulated, orbit tracking being an example, tests and performance measurements associated with the on board TTC subsystem are made. These tests demonstrate operability and establish performance characteristics that are necessary for post-launch tracking activities.

3.5 PRE-LAUNCH EARTH SEGMENT

During the constructional phases of a spacecraft, support equipment and services will be required to enable test activities to be performed. The same principles which will be described in this section can be equally applicable to spacecraft that are to be engaged upon scientific missions.

The initial stages of construction for the application satellites entails the building of units. This is followed by the integration of the units into subsystems. The subsystems are then combined upon a mechanical structure to form the satellite itself. For all these activities, including launch operations, support equipment will be required. This equipment, which is generally termed ground support equipment (GSE), consists of two prime categories; mechanical ground support equipment (MGSE) and electrical ground support equipment (EGSE).

3.5.1 Mechanical ground support equipment (MGSE)

The MGSE is necessary so that the satellite can be transported and manoeuvred during the construction and testing processes. The most significant factor regarding MGSE designs is that the mechanical adaptive interface for the satellite and the launch vehicle attachments are almost an exact duplication of the expected fitting. Therefore, to accommodate vibration test requirements, an MGSE vibration adaptor is employed to mount the satellite to the vibration facilities. Similarly, an MGSE thermal adaptor is employed for the thermal vacuum test. The MGSE thermal adaptor enables the satellite to be suspended and manoeuvred in the test chambers so that a representative simulation of the orbital environment can be achieved. Other MGSE components are used during satellite level AIT, and also to cope with mounting the satellite in the transportation containers utilized during AIT for movements between different test sites.

3.5.2 Electrical ground support equipment (EGSE)

The EGSE comprises the electrical test equipments that are required during the production of a satellite. For communication satellites that have been constructed under the auspices of ESA the EGSE is required during three distinct levels of satellite construction:

- unit level,
- subsystem level,
- system level.

    The EGSE is therefore needed to support all satellite construction operations, commencing with unit level production and concluding with the final countdown of the launch vehicle on the day of the launch.

    The EGSE is the Earth segment that operates and tests the satellite before its launch. Thus, the EGSE must be able to control the satellite by imitating the post-launch Earth segment. The EGSE should also enable some aspects of the space environment to be simulated. Therefore, the pre-launch Earth Segment, the EGSE, must provide facilities to:

- receive and process the housekeeping telemetry that is transmitted from the satellite,
- encode and transmit orders to configure the satellite by telecommand,
- provide electrical power to the satellite,
- stimulate on board units with signals that will be representative of those which will be experienced during orbital operations,
- perform measurements and functional tests of on board subsystems.

    The EGSE has the capability to mimic the Earth segment which will be used during all post-launch satellite operations. Thus, the EGSE performs similar functions to the post-launch Earth segment. In addition, the EGSE must also complement the environmental test facilities to enable and support satellite testing. The pre-launch EGSE requirements can be summarized as being tools which provide facilities for the construction, testing, and operation of the satellite.

    For pre-launch satellite system level activities, the EGSE is generally deemed to be the checkout equipment or ground checkout station. This is because most EGSE elements have the capability to stimulate and measure during testing operations and therefore perform complete tests. A complete test is a check of both the satellite's operation and performance, that is a complete checkout of satellite functions. These tasks can be and often are performed simultaneously by the EGSE.

    The EGSE architecture embraces satellite subsystem specific checkout equipment (SCOE) and the overall checkout equipment (OCOE) which controls the operation of the SCOEs. The OCOE also initiates satellite telecommands and processes satellite telemetry data, thus exercising overall control of satellite operation and test. An overview of this satellite system level EGSE architecture is given in Fig. 3.6. A comparison with Fig 3.5 will reveal some basic commonalities that exist for pre- and post-launch operation.

    The term overall for the OCOE is somewhat of a misnomer, as is the term equipment in this case. These facts can be attributed partly to historical reasons. In the early days of European satellite operations when these tasks were not so complex, equipment which undertook the overall test and performance activities were designated as the OCOE. Within modern EGSE architectures, the OCOE is the executive master controller of the EGSE, the pre-launch Earth segment.

    The SCOEs are controlled from the OCOE over a data multiplex system, and are

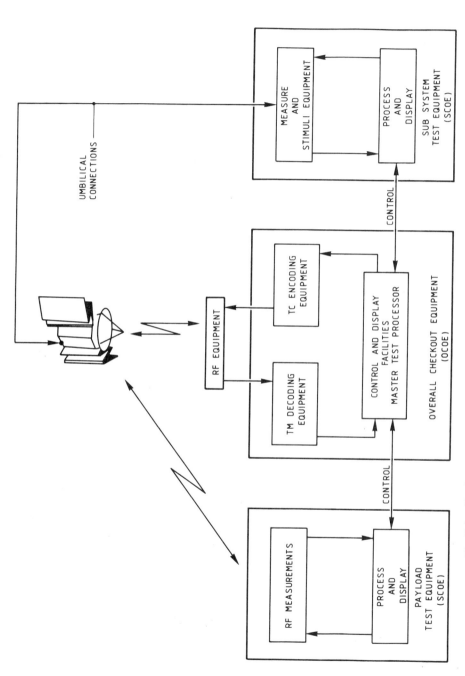

Fig. 3.6 — Earth segment for pre-launch operations. The OCOE operates the satellite so that tests can be undertaken via RF and umbilical connections.

connected to the satellite by an umbilical connection comprising cabling and test connectors. The umbilical connection allows test signals to be injected into the satellite and enables special test points (which are not accessed by the on board satellite telemetry) to be monitored. The power supply subsystem (PSS) SCOE facilitates the testing of the on board power supply system and provides power to the satellite during testing. The attitude and orbital control (AOCS) SCOE tests the on board subsystem, and the payload SCOE tests the onboard communication payloads.

The OCOE computer, designated the master test processor (MTP), operates the system in an executive capacity by means of user software. This provides centralized control and display facilities from the EGSE executive control workstation which is the test conductor's console (TCC). Other terminal equipment, known as observer consoles, allow several engineers to participate in the testing operations. These consoles are EGSE workstations over which the TCC has the master executive capability.

The test conductor engineer can initiate a telecommand at the TCC, resulting in a telecommand transmission to the satellite under test. A telemetry response is received at the OCOE, processed by the MTP and the result relayed to the TCC. Similarly, the test conductor can initiate a stimuli signal from a SCOE to the satellite from the TCC. Such action results in an order being transmitted from the MTP over the data multiplex system to the SCOE concerned. The SCOE then generates the signal which is injected into the satellite through the umbilical test connector. The response can be monitored via satellite telemetry at the OCOE. In addition, a special test point within the satellite may allow the response to be received by the SCOE. In this case, the SCOE may analyze the response and transmit the result over the data multiplex system to the OCOE. The test conductor, being the executive controller of the test activities, is therefore able to monitor the results of both these actions. During these activities, the OCOE performs archiving tasks so that a history file of test data can be compiled. This enables an analysis and comparisons between test results to be performed at a later time.

This EGSE architecture or a derivative configuration is utilized during all pre-launch system level activities. Such configurations are employed at the launch site and give a satellite GO/NOGO decision that is a vital factor for the launch vehicle countdown operations. From a systems engineering standpoint, the test equipment (the EGSE) may be more complex than the item under test (the satellite) to facilitate intensive testing.

The OCOE user software, which operates with the most advanced European OCOEs, is a version of the European test and operations language (ETOL). It is ETOL which provides the user friendly man–machine interfaces at the OCOE workstations. The majority of spacecraft which have been built in Europe have utilized a version of the EGSE system level test architecture that is depicted in Fig. 3.6, and ETOL. The language is described in some detail in Chapter 4, and is the subject of the Appendix.

3.5.3 Launch support configurations

Before the launch of the communication satellite, a simplification of the EGSE configuration is made, but the executive control of the satellite is maintained by the

OCOE. One change which is made during the final pre-launch activities is that the communication payload SCOE becomes redundant as the payloads are not normally powered. Another modification to the EGSE is that the AOCS SCOE is not required, because the attitude and orbital control subsystem will have undergone final functional tests. The prime connections that are made to the satellite are for the supply of electrical power to keep the on board batteries fully charged. In addition, telemetry and telecommand liaisons are maintained under OCOE control, enabling satellite health to be monitored. These connections would enable the satellite to be adequately controlled should a launch vehicle countdown be aborted. The connections for the power supply and satellite control support services are made via a launch vehicle umbilical connection.

The EGSE configuration for the final pre-launch operations is functionally the same as that which will be utilized during post-launch operations. This means that the pre- and post-launch Earth segment configurations for satellite control are similar, the most significant difference being the distance between telemetry and telecommand support equipment and the satellite for the pre- and post-launch situations. Another difference is that the spacecraft is powered by the PSS SCOE before launch but in space it will be powered by the solar arrays being illuminated by the Sun.

For the final activities associated with the launch, the prime satellite support facilities are the EGSE items, power supply subsystem specific checkout equipment (PSS-SCOE) and the overall checkout equipment (OCOE). The PSS-SCOE operates under the executive control of the OCOE which has the master control of satellite pre-launch operations. This is the basic configuration that is employed during the final phase of launch vehicle countdown activities. This final phase commences a few hours before launch vehicle engine ignition and terminates very shortly after this event, when the launch vehicle commences its flight.

Support equipment will be necessary for the control and operation of the expendable launch vehicle (ELV). The ELV support equipment is fully operational during the vehicle countdown, engine ignition, and flight, with activities terminating at the end of the launch trajectory.

For an Ariane 4 launch, the ELV electrical ground support equipment, termed ground stations, are employed at several locations. This includes those which are close to the launch sites in Kourou, French Guiana, and ELV ground stations located at sites in Brazil, Ascension Island and in Gabon. These stations monitor and control the launch and flight of Ariane during this crucial phase of space segment construction. These details are given in Figs 3.7 and 3.8, which take a dual launch as an example. The station at Libreville monitors the injection into orbit of the Ariane third stage, together with its payload, the two spacecraft. This station continues to monitor the final and equally important phases of Ariane operations.

After successful launch operations, the satellites will come under the control of the Earth segment facilities which support mission control: the ground segment (OCC and TTC stations) of the post-launch Earth segment architecture.

3.5.4 A pre-launch verification

If the ground segment to be used after satellite launch is different from that used during pre-launch activities, it should be verified to establish its suitability for utilization.

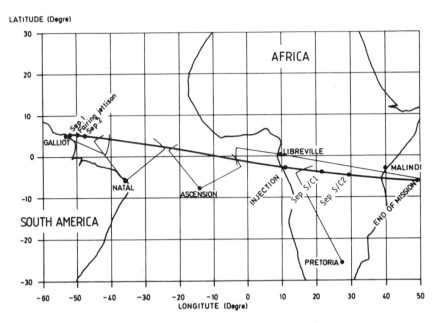

Fig. 3.7 — Ariane 4 ground station visibility. Sep 1 — Separation of first stage; Sep 2 — Separation of second stage; Sep S/C1 — Separation of spacecraft No. 1; Sep S/C2 — Separation of spacecraft No. 2. Ground station location for critical activities on Ariane flight path (courtesy of Arianespace).

For many of the communication satellites which have been constructed in Europe, it is a normal requirement that the post-launch ground segment undergoes a verification test programme. This is a test and validation activity to demonstrate that the facilities which will be utilized for satellite orbital operation and control are suitably developed before the satellite's launch. This requirement necessitates that the satellite is controlled before launch by the ground systems. This is undertaken to demonstrate the ability of the ground systems to perform these functions in a satisfactory manner. These tests are a significant part of the satellite pre-launch preparations.

It has been explained previously that the ground support facilities of the post-launch Earth segment for control of communication satellites are normally sectioned into the operations and control centre (OCC) and the telemetry, tracking, and command (TTC) stations. The TTC stations provide the RF interface to the satellite; the OCC, via computer software, directs and manages satellite operations and control.

TTC station compatibility testing for the majority of European communication satellites has been a verification of the inter-segment RF interface. This has included performance tests which have not necessitated the use of a satellite model. Thus, this TTC station verification task has no impact upon satellite construction schedules.

The OCC computer software tested is that which will be used during the in-flight life of the satellite. The satellite to be launched is employed for these tests.

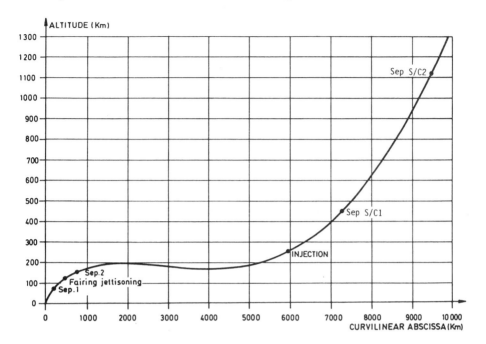

Fig. 3.8 — Ariane 4 nominal flight profile. Sep 1 — Separation of first stage; Sep 2 — Separation of second stage; Sep S/C1 — Separation of spacecraft No. 1; Sep S/C2 — Separation of spacecraft No. 2. Altitudes at which the most critical activities occur (courtesy of Arianespace).

Therefore, there can be impacts upon satellite construction schedules to enable this OCC software to be validated.

The Earth segment configuration for the OCC software validation tests utilizes the major items of electrical ground support equipment (EGSE) which are used during satellite construction. This Earth segment comprises overall checkout equipment (OCOE) and some SCOE items. The OCOE is the equivalent of the OCC and TTC stations for telemetry and telecommand purposes. Compatibility testing for the OCC software can be conducted with the satellite being remotely located from the OCC. This is achieved by using telephone lines as the communication links for the telecommand (TC) signals from the OCC to the satellite, and telemetry (TM) signals from the satellite to the OCC. The basic system configuration for the OCC computer software validation tests is shown in Fig. 3.9. An OCOE ground station adaptor (GSA) is an additional small item of electrical ground support equipment required to enable the OCC to be interfaced to the satellite via telephone lines, modems, and the OCOE TM and TC equipment.

The OCC interfaces with the GSA are made with signals that are in the form of a synchronous binary stream which is pulse code modulated (PCM) in split phase level (SPL) form. The input to the GSA for transmission to the OCC is also in the form of a PCM-SPL stream from the OCOE TM decoding equipment. More details and explanations of PCM telemetry and telecommand signals and standards are given in Chapter 4. These standards are a prime requirement for satellite post-launch

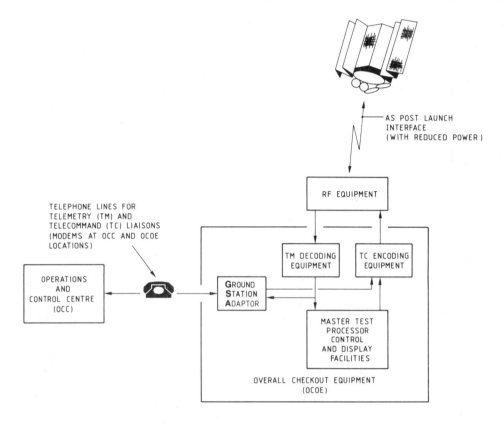

AS POST LAUNCH
INTERFACE
(WITH REDUCED POWER)

RF EQUIPMENT

TELEPHONE LINES FOR
TELEMETRY (TM) AND
TELECOMMAND (TC) LIAISONS
(MODEMS AT OCC AND OCOE
LOCATIONS)

TM DECODING
EQUIPMENT

TC ENCODING
EQUIPMENT

OPERATIONS
AND
CONTROL CENTRE
(OCC)

GROUND
STATION
ADAPTOR

MASTER TEST
PROCESSOR
CONTROL
AND DISPLAY
FACILITIES

OVERALL CHECKOUT EQUIPMENT
(OCOE)

Fig. 3.9 — Configuration for OCC software validation tests. This basic configuration of ground segment elements enables the software to be validated before launch without a satellite simulator.

operations. The interface between the OCOE and the satellite does conform to these standards, but the RF interface links are made with reductions to the RF powers that are necessary for post-launch liaisons.

The initial validations of the OCC software have been conducted with the satellite at a European location. However, it can be necessary for a final verification to be undertaken when the satellite is at the launch site and undergoing final preparations on the launch vehicle. For example, the final OCC software verification for ESA's communication satellite OTS (orbital test satellite) was conducted during launch preparations at the Kennedy Space Centre in Florida, USA. Similarly, some of the ESA European communication satellites (ECS) and maritime European communication satellites (MARECS) have had their OCC software finally verified during launch preparations at the Agency's launch site in French Guiana, South America.

The OCOE retains executive control of satellite operations for OCC compatibility testing. This control is, of course, delegated by the OCOE to the OCC for the validation tests to be performed. Successful completion of these tests will demonstrate the OCC's ability to be assigned the executive control of satellite operations after launch. Successful completion of OCC software verification testing has been a prime requirement for the go ahead for launch preparation to be given.

To consider more fully the details of rationalizing the pre- and post-launch Earth segment, the interfaces within and external to an aerospace system need to be examined.

Interface definitions are of a prime importance for the implementation of good systems engineering techniques. An interface change within a system can, as demonstrated, be accommodated, but consideration does require a long term, open-minded approach. The interfaces associated with the control of an aerospace system are examined in Chapter 4.

4

System control:
interfaces and configurations

An aerospace system comprises elements which are located within both space and Earth segments. Information traffic between these two segments can be divided into two types:

(a) Information required to monitor and control the health and configuration of the on board spacecraft subsystems.
(b) Information from the spacecraft payloads to the ultimate users.

Both types of information require interface equipment within space and Earth segments. This chapter describes the interfaces that are necessary to fulfill the requirements of (a).

The inter-segment control interface which satisfies these requirements is defined by the telemetry, tracking and command (TTC) specifications and standards. The on-going developments of spacecraft technologies will require that inter-segment control interfaces evolve to meet expanding requirements. However, in this chapter the established ESA PCM telemetry and PCM telecommand standards will be used as an explanatory medium. These standards are referenced in the *Bibliography*.

Some inter-segment liaisons for the subject communication system will not need to perform according to the TTC standards. The user Earth stations do not require TTC liaisons to monitor and control satellite health. The user Earth stations are part of the communications link. The satellite health control tasks, explained in Chapter 3, are performed under the executive control of the ground segment. Therefore, in a communication system, the ultimate user will not be directly engaged in satellite control activities. However, the ultimate user of scientific experiments may be delegated some direct control of a spacecraft. An overview of the interfaces associated with an aerospace system is given in Fig. 4.1. It will be observed that other important interfaces are shown in this figure beside the inter-segment control interface.

Aerospace systems will function with interfaces which are internal and external to

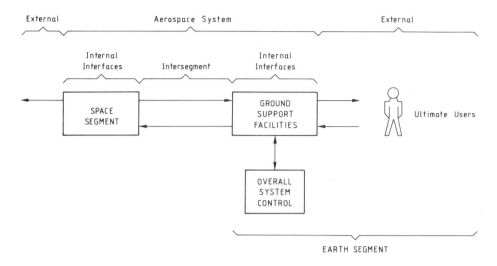

Fig. 4.1 — System interfaces : an overview. The interfaces between elements of an aerospace
system are shown, together with those which can be regarded as external.

the space and ground segments. For the communication system, there is no true
external interface for the space segment which serves the direct requirements of the
ultimate users. Their direct requirement will be to communicate with each other,
using communication links of which the satellite payload is a part. This exploitation
of space uses the location of the geostationary satellite to achieve an efficient
communication link. The external interfaces into space will be associated with
satellite control. These interfaces relate primarily to the satellite's attitude and
orbital control subsystem (AOCS) and its thermal subsystem. These play a major
role in maintaining the operability of the satellite. In particular, Earth, Sun, and star
sensors of the AOCS fall into this external interface control category.

When the aerospace system has a spacecraft which is engaged upon a scientific
mission, the external space segment interface will be employed to meet spacecraft
engineering requirements which, in principle, are similar to the communication
satellite. External space segment interfaces that perform an exploratory task can use
the probing instrumentation on board a spacecraft. In such a case, this external space
segment is not only being employed to meet engineering requirements but also to
satisfy scientific aims. At the other extreme of this aerospace system is the scientist
who will analyze the results of the scientific probe.

Interfaces associated with space segments can exhibit significant commonalities
for both space exploitation and exploration endeavours. These commonalities are
centred around spacecraft health using the inter-segment control interfaces.

The basic functional architecture of an aerospace system that can be applicable to
both scientific and application spacecraft is shown in Fig. 4.2. Closed-loop operation
of the telemetry (TM) and telecommand (TC) interfaces can be made via the
executive facilities. These employ workstations to provide the man–machine inter-

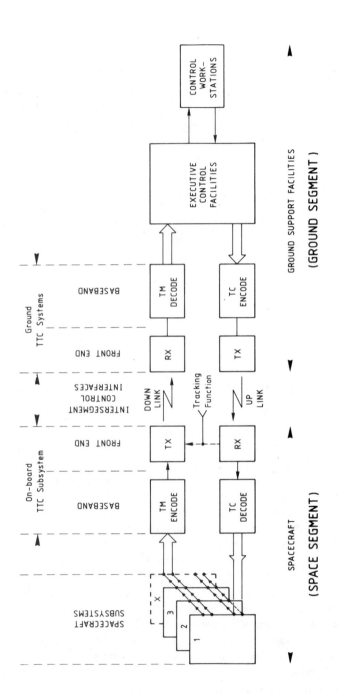

Fig. 4.2 — Aerospace system functional architecture. The elements and interfaces that enable the system to function and satisfy user requirements are shown.

face (MMI) for overall system control. The MMI enables the inter-segment interfaces to be used so that the aerospace system can be operated efficiently to meet user requirements.

4.1 INTER-SEGMENT CONTROL INTERFACES

The inter-segment control interfaces which are necessary for safe and efficient system operation are provided by the TTC facilities. Within the Earth segment, this support is made with equipments and systems that, as explained previously, are termed the ground segment. The TTC interfaces are of prime importance from an overall system control aspect as they provide the possibility for closed-loop control between the space and ground segments.

There are significant similarities inherent in telemetry, tracking and command technologies. Telemetry is a transmission from space to ground segment and is designated the downlink; telecommand transmissions are in the opposite direction and are termed uplink. From the explanations given in earlier chapters, it is implicit that the tracking function can incorporate the application of uplinks and downlinks.

Definitions which are fundamental and common to telemetry and telecommand signals are shown in Fig. 4.2 and are described in the following sections.

4.1.1 Front-end equipment

The uplinks and downlinks for TTC control are established with the front-end equipment. This equipment provides the capability for radio frequency (RF) transmissions and receptions between the space segment and ground segment for telemetry and telecommand signals. The equipment incorporates antennae, receivers (Rx) and transmitters (Tx) as a minimum, and normally enables the tracking function to be performed.

4.1.2 Baseband equipment

There are further inherent similarities in telemetry and telecommand principles apparent in the baseband equipment.

This equipment incorporates the digital technologies for handling the telemetry and telecommand signals. The baseband equipment is used to connect the front-end equipments with the executive controller within the ground segment. For the space segment, the baseband equipment interfaces the telecommand and telemetry signals of the on board systems with the space segment front end.

4.1.3 Technical terms

The following technical terms are common and are relevant to the baseband equipment concerned with both the telemetry and telecommand subsystems. These terms comply with and are covered in more detail in the ESA PCM telemetry and telecommand standards referenced in the *Bibliography*.

- video signal: This is a common definition for the baseband serial signals.
- PCM : Pulse code modulation. Utilizes the basic properties of digital logic circuits. The codes which are most generally used are NRZ-L (non return to zero-level) and SPL (split phase level).

- clock signal: A level change which is synchronous with the PCM signal. The clock rate for an NRZ-L is equal to the bit rate and twice the bit rate for SPL coded signals.
- bit : Binary digit. A logical '1' or '0' which occupies one time period in a PCM data stream.
- byte/octet : Eight contiguous bits. An eight bit word.
- PSK : Phase shift keying. Utilized for sub-carrier modulation.

Fig. 4.3 gives a pictorial presentation of these terms. Further details regarding telemetry and telecommand interfaces which are guided by the relevant ESA PCM standards are given in the following sections.

4.2 TELEMETRY INTERFACE FUNCTIONS

Telemetry (TM) is the term given to the information transmitted from the space segment to the Earth segment. The telemetry information comprises parameters relating to on board spacecraft measurements and status. This information, therefore, can represent analogue and digital parameters.

In the case of a system which includes a scientific probe in space, two types of information are transmitted from the space segment to the Earth segment. One type of information will be the data which have been gathered by the probing instruments. These data will be scientific information which needs to be relayed to Earth from the spacecraft for the scientific user. To do this, it will be encoded and digitized so that it can be transmitted to the Earth segment as a telemetry signal. When this signal is received, it will need to be decoded and presented to the scientist. Special computer software will be required to perform this task. Ideally, this software should provide a user friendly interface which will enable a man–machine interface to be established with a readily understandable dialogue. The second type of information required from the scientific spacecraft is that associated with the control of its status and health. This information is of an engineering nature and will include analogue values of current, voltage, and temperature and digital values concerning the ON/OFF and prime/redundant status of on board equipments. This analogue and digital information is transmitted as telemetry data and is designated housekeeping telemetry.

For the subject communication satellite, the telemetry communications will be entirely for housekeeping purposes dealing with satellite health control. This control task is undertaken by the ground support facilities (ground segment) of the post-launch Earth segment.

4.2.1 Telemetry format

On-board spacecraft information is encoded into telemetry messages. Each individual telemetry message is generally known as a format. A format comprises parameter samples which are taken over a defined period. The telemetry parameters within the format are time-division multiplexed (TDM). The formats are assembled by the on board space segment telemetry encoder into a PCM signal which after phase shift keying (PSK) modulation, forms the video signal for the input to the RF transmitter. The telemetry data transmitted across the inter-segment interface are in the form of a continuous serial data stream. When the telemetry signal is received by

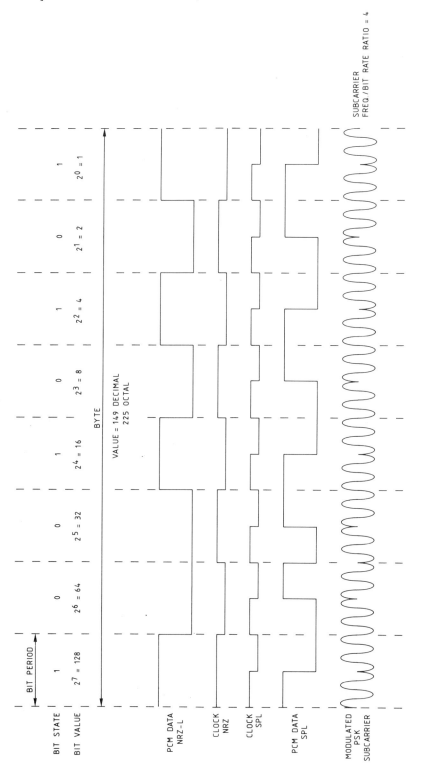

Fig. 4.3 — Video signal characteristics. Relationship between signals of PCM data stream.

the ground segment it will be required to be decoded and processed. Fig. 4.4 shows the equipment concerned with the encoding, reception, and decoding of telemetry messages in a simple functional form.

The formatting of the data stream is the task of the on board pulse coded modulation (PCM) telemetry encoder. The basic functions of the PCM TM encoder are:

- The sequential sampling, of all encoder inputs. These inputs are parameter samples relevant to each on board subsystem. Parameter samples are termed source data.
- Conversion of the encoder inputs into telemetry data. The telemetry data are in digital form, thus conversions are necessary for analogue parameters.
- Encoding the TM data into a serial message in the form of a PCM signal.

The telemetry data are transmitted in the form of an uninterrupted data stream, comprising structured messages termed formats.

4.2.2 Telemetry message details
A typical TM format relevant to the ESA PCM telemetry standard is given in Fig. 4.5, the following definitions are applicable.

- Format : an integral number of TM frames.
- Frame : an integral number of consecutive TM words.
- Word : 8 consecutive binary bits, designated a byte.
- Channel : a TM channel is an input to the on board TM encoder. In the TM format, channels are represented as words or parts of words thus:
  ★ one data sample, as in the case of analogue channels or 8-bit serial channels.
  ★ a part of one data sample, as in the case of 16-bit serial digital channels or parallel digital channels (64 bits).
  ★ 8 data samples, as in the case of bi-level (one or zero) channels.

Some words in the format are allocated to channels that are not telemetry parameters. That is to say, they do not represent either analogue or digital values from sensors or status registers. These channels are defined as follows and are depicted in Fig. 4.5.

- Frame synchronization channels (FRAME SYNC)
  A fixed bit pattern repeated at the beginning of each frame. It is a form of identifier that assists the decoding process performed by the ground segment.
- Spacecraft identification channel (SAT ID)
  A fixed bit pattern that identifies the spacecraft which is transmitting the telemetry data. This allows the unique identification of the spacecraft to be determined by the ground segment.
- Frame counter channels (FRAME ID)
  The frame counter counts up from the first frame to the last in the format. In Fig. 4.5, the format is composed of 12 frames, thus the frame counter cycles from 0 to

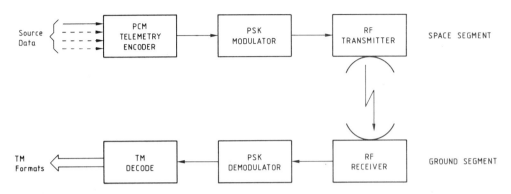

Fig. 4.4 — Telemetry system functional architecture.

11. These channels are often termed the frame identification (ID) channels. They are used in the ground segment decoding process.
- Format counter channel (FORMAT ID)
  This channel is used to transmit the contents of a counter which is incremented every format. Therefore, formats have an address so that individual formats can be identified. This channel does reset eventually, depending upon the size of the on board format counter.
- Spacecraft timing channel.
  This channel is allocated to the contents of a spacecraft TTC subsystem timing counter. This counter has a better resolution than the format counter, and can, therefore, be utilized for on board spacecraft event timing.

The frame synchronization channels enable the ground segment to be suitably synchronized with the received telemetry formats, and in conjunction with the frame and format counter channels enables decoding and processing to be accomplished.

4.2.3   Channel allocation
The various spacecraft subsystem parameters, for example temperatures, pressures, and voltages, are allotted channels which have definitive locations in the TM format. The TM format which complies with the relevant ESA standard allows the sampling and encoding of data in a regular cyclic manner. This means that once allotted a channel address, the measured source data variable will always appear in the same place in the TM format. Examples of channel addressing are shown in Fig. 4.5 where it should be noted that addressing is normally assigned in octal notation.

4.2.4   Telemetry data sampling rates
The sampling rates of channels can be different, and are identified on Fig. 4.5.

Frame Commutated : once per frame
Super Commutated : More than once per frame

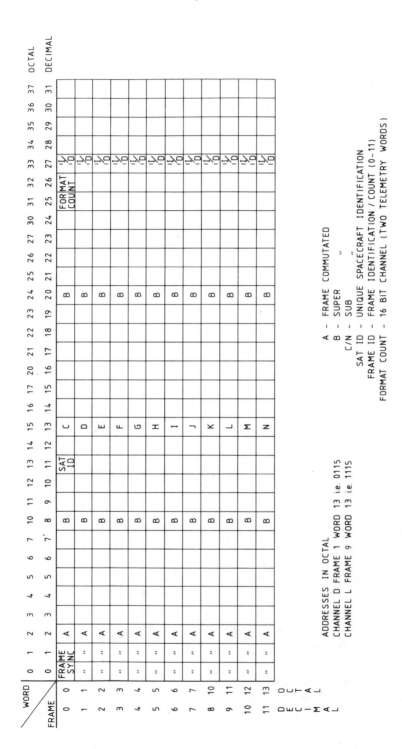

Fig. 4.5 — Telemetry message structure —TM format.

Sub-Commutated   : less than once per frame, but at least once per format.

The term commutation is used as it originates from the operational principles of the PCM encoder which is analogous to the function of a commutator.

4.2.5   Telemetry data transmission

The format is transmitted by the spacecraft as a serial data stream. Word zero of frame zero is the first word transmitted, followed by word one, two,and so on of frame zero. Word thirty one (decimal) of frame zero is followed by word zero of frame one. This continues to the end of the format. The next format will then be transmitted without any message interruption. The frame identification (ID) will start at a count of zero again and the format count will have been incremented by one. The frame ID and format count are checked and utilized by the ground segment operational software for data processing.

4.3   TELECOMMAND INTERFACE FUNCTIONS

One of the inter-segment control interface functions will be the transmission of commands from the Earth to the space segment to configure and control the spacecraft.

Telecommand (TC) is the term used to refer to the orders and instructions which are encoded and transmitted. The encoding process results in the formatting of a telecommand message with data added for validity detection purposes. The description of the telecommand interface which follows complies with the ESA PCM telecommand standard (see *Bibliography*).

Telecommand messages are constructed in frames which after decoding and distribution on board the space segment, result in some action being effected. When an action of a telecommand order is completed, the term telecommand execution is often employed. The equipment concerned with the encoding, transmission, reception, decoding, and distributing of telecommands is depicted in functional form in Fig. 4.6. The structures of a telecommand message and telecommand frames are shown in Fig. 4.7, and these are explained in the following text.

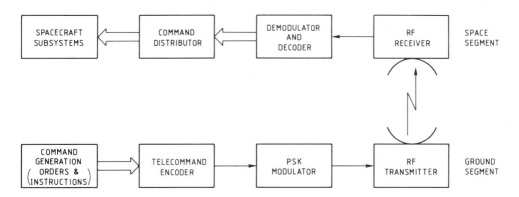

Fig. 4.6 — Telecommand system functional architecture.

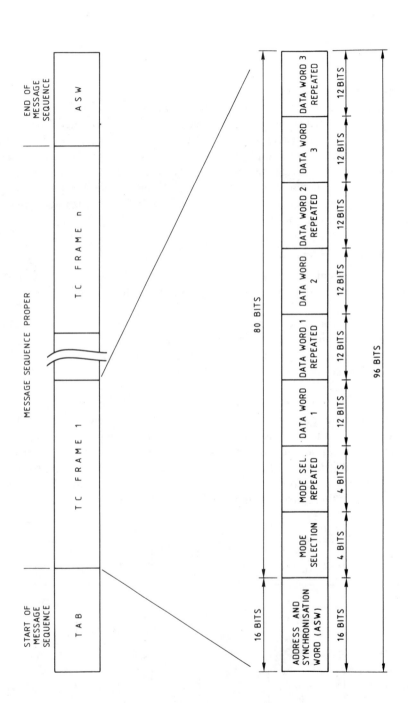

Fig. 4.7 — Telecommand message structure.

4.3.1   Telecommand format

Telecommand orders and instructions are encoded and formatted into a frame which is part of a telecommand message. A telecommand message can consist of multiple frames. Only parts of the telecommand frame enable an action on board the spacecraft to be initiated. All the other bits in the message are concerned with the acquisition and correct decoding of a telecommand. These bits are inserted to ensure that the message is received and decoded only by the addressed spacecraft decoder and to enable errors within the message to be detected. If errors are detected, the telecommand message is rejected and re-transmission is instigated so that the required actions can be undertaken.

The telecommand message is formatted, pulse code modulated, superimposed on a PSK sub-carrier, and sent as a video, baseband signal to the RF transmitter stage. The hardware on board the spacecraft receives and decodes the message, and distributes the order and data to various on board satellite subsystems. Telecommands are normally transmitted only when required, therefore the telecommand RF carrier is not usually subjected to permanent transmission.

4.3.2   Telecommand message details

The following is a brief description of the components of a telecommand message shown in Fig. 4.7.

- Start of message sequence. This consists of a fixed number of timing and acquisition bits (TAB). These are used to initialize the message and allow the decommutation process to start within the on board spacecraft TTC subsystem.
- Message sequence proper. A telecommand message consists of an uninterrupted sequence of telecommand frames which commence with the spacecraft address and synchronization word (ASW).
- End of message sequence. The spacecraft ASW is also used to terminate the message.

Fig. 4.7 shows that the majority of telecommand frame components are repeated. This is to allow checks to be made so that commands can be rejected if errors are detected and corrective actions undertaken.

4.3.3   Telecommand frame

The specific features of a telecommand frame are now identified and described in the sequence transmitted.

(a) Address & synchronization word (ASW). This is a 16-bit word assigned to each spacecraft telecommand decoder for identification purposes and to indicate the start of a telecommand frame. It is also utilized to terminate telecommand messages.

(b) Mode selection. This is a 4-bit word which follows the ASW. The mode selection word allows command data to be correctly distributed to and utilized by the on board subsystems.

(c) Data word. There are three command data words, each comprising 8 bits of user data. To cope with errors each word is followed by 4 check bits which are assembled according to a Boolean algorithm applied to the individual 8 data bits. This enables a further check of command data to be made. Rejections occur if errors are found.

4.3.4 Telecommand modes

Although the telecommand frame is always structured in the same manner, the use of the 24 user data bits (data word 1, 2, and 3) depends on the telecommand mode. There are three basic types of telecommand operations which are dependent on the mode selection for the on board command distributor unit.

The design of the command distributor unit can allow it to operate as follows in accordance with the mode selected:

- Distribute the 24 bits of data words 1, 2, and 3 to a defined location.
- Distribute the 16 bits of data words 2 and 3 to an address which is defined by data word 1.
- Distribute single pulses to locations which are addressed by data words 1, 2, and 3. The signals produced drive relays or electronic circuits and are commonly known as direct ON/OFF commands.

Direct ON/OFF telecommands are executed immediately after reception and decommutation. The term telecommand execution is often used for the result of telecommand decommutation and the corresponding action on board the spacecraft.

4.3.5 Telecommand on board timing

A facility can be provided that enables a telecommand to be executed at a predefined time after its reception on board the spacecraft. This is termed a time tagged telecommand. The telecommand message contains an encoded time at which the command is to be executed on board the spacecraft. This type of command operation makes use of one of several registers within the TTC subsystem which are known as a time-tag registers. Each time-tagged telecommand is allocated a register.

A time-tagged telecommand message is composed of two adjacent telecommand frames. The first frame of the message contains the address of the time-tag register and the data giving the spacecraft time at which the command should be executed. This time is loaded into the addressed time-tagged register. The following frame contains the actual telecommand, that is the order or instruction which is required to be executed. The telecommand is executed when the time in the time-tag register is identical to that in a register which contains the incrementing spacecraft time. This facility allows a delay between the reception and decommutation of the telecommand signal and its execution. Therefore, on board actions can be programmed to be performed without the need for ground segment contacts at the time when the command should be executed.

4.4 TRACKING INTERFACE FUNCTION

In addition to the telemetry and telecommand facilities required for overall system control, there will also be the requirement for ground support equipments to provide tracking facilities so that the in-flight position of the spacecraft can be determined. For scientific probes and geostationary application spacecraft, the detailed technical requirements for the tracking equipment are normally different. This is a direct result of the required post-launch flight paths.

The purpose of the tracking function is to follow the flight path of a spacecraft. This enables a check to be made against the flight path predictions, thus facilitating any corrective actions to be taken. Furthermore, the tracking function should acquire and maintain good RF connections for telemetry and telecommand transmissions between the Earth segment and the space segment. There are several techniques which can be employed to enable this function to be performed.

If tracking information is collected from several ground segment locations, an orbital flight path can be determined. For a geostationary communication satellite this function will require the facilities of several TTC stations during the elliptical equatorial transfer orbits. The transfer orbit is determined and monitored before the apogee boost motor is fired to attain the required geostationary position for the satellite. After this has been achieved, a single TTC station can be employed, and the tracking function becomes a satellite health monitoring task.

Perturbations in the orbit do occur. Environmental factors which influence the nominal orbital path of the satellite include:

- The attraction of the Sun, Moon and the Earth.
- Solar radiation pressure (solar wind).

The determination of an orbital position and path of a spacecraft can involve three prime measurements that are made from TTC stations. These are:

(a) The distance of the spacecraft from the TTC station, the range, which is made simultaneously with (b)
(b) the angular position of the spacecraft relative to the TTC station axes. This involves the determination of azimuth and elevation angles in relation to the geographical location of the TTC station.
(c) The rate at which the spacecraft to TTC station distance is changing — the range rate.

The distance between a spacecraft and a TTC station can be determined by measuring the phase difference between a signal transmitted to a spacecraft and the same signal re-transmitted from the spacecraft. This task can be performed by a TTC station superimposing signals upon the telecommand RF carrier signal. These signals are received on board the spacecraft and can be re-transmitted on the RF telemetry carrier signal. The phase difference between transmitted and received signal is a measurement of the elapsed time between transmission and reception. The elapsed time is proportional to the distance (the range) of the spacecraft from the TTC station performing the tracking task. The on board spacecraft TTC subsystem imparts a delay between signal reception and retransmission. This delay value is established during pre-launch testing and is taken into account during the orbit determination process.

The orbit determination process requires the angular position of the spacecraft relative to the TTC station to be established. The simplest method for the determination of the azimuth and elevation angles is to move the TTC antenna and search for maximum received signal, this is often termed autotracking. The principal purpose of autotracking is to move the antenna to maintain maximum signal strengths for the telemetry and telecommands received from and transmitted to the spacecraft. Continuous automatic tracking of a spacecraft utilizing the received signal can be performed.

The most common tracking method is known as step-by-step tracking. The TTC antenna is moved in both axes, azimuth and elevation, in a series of steps, and the signal levels for each of the antenna positions are compared. The results of the comparisons are used to determine the antenna angles of pointing for maximum signal strength.

When a geostationary satellite is on-station, range rate measurements are not normally taken. However, they can be made during transfer orbit determination by monitoring the doppler shifts of the tracking frequencies.

4.4.1 Station keeping
A geostationary satellite needs to be maintained in a specific position in its orbital path relative to the centre of the Earth. To keep the satellite in its optimum orbital position it may be necessary to make some corrections at regular intervals. Because the satellite is subject to various disturbances, a continuous geostationary position is not attained; in fact the satellite will drift from its optimum position. Consequently, the satellite is maintained between limits which are often termed a window. The limits of the window are defined by angles around the optimum position. Station keeping is the term which is given to that activity which keeps the satellite within this window whilst utilizing as little of the on board fuel as possible. Fig. 4.8 gives an overview of an operational window. Maintaining the spacecraft within the window is performed by manoeuvres that are generally termed north–south and east–west station keeping which are corrections for drifts in these directions.

To accomplish the station keeping task, it is required to track the satellite so that control of its position can be maintained. The tracking information together with data which are telemetered from the attitude and orbital control subsystem (AOCS) of the satellite, is used to correct orbital and attitude deviations. The corrections can be implemented by means of telecommands or by automatic actions on board the satellite if dangerous conditions develop. Such activities are performed by the operation of AOCS components enabling orbital deviations to be corrected by the north–south and east–west manoeuvres. Typically, north–south corrections are made at monthly intervals and east–west corrections are undertaken twice during this period.

4.5 EXECUTIVE CONTROL

The telemetry, tracking, and telecommand (TTC) interfaces that have been described need to be operated and controlled under executive control. For safe and efficient operation, executive control facilities will have two prime external interfaces. One of these interfaces will be associated with the ground to space segment

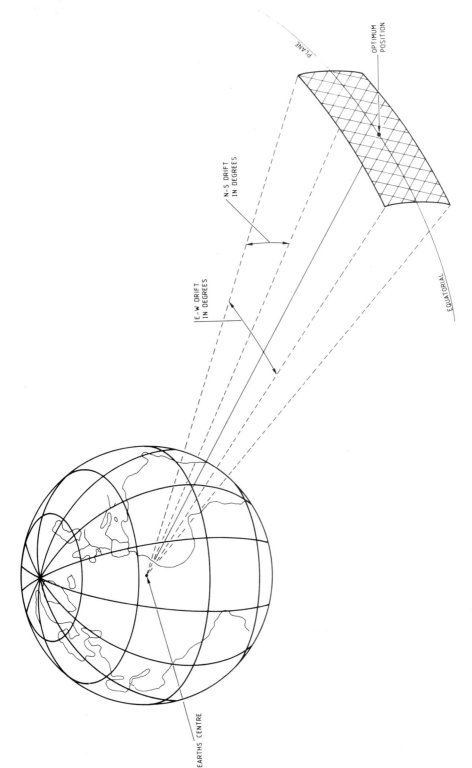

Fig. 4.8 — Station keeping window. Permissible limits of drift from the optimum geostationary position.

TTC links. The other interface will be that which provides the liaisons between the executive control machine and its operators, that is a man–machine interface.

The man–machine interface at the spacecraft control workstations can be similar for spacecraft which are engaged upon scientific probes and those which exploit space. System performance will depend greatly upon the correct operation and control of the space segment. The personnel associated with the real-time task of in-orbit spacecraft health control are located at what are termed mission control workstations. The term spacecraft controller is that which is often assigned to the senior engineer who is located at one of these workstations. When a spacecraft is fully operational and satisfying all the requirements of the overall system, the activities at these locations can be of a predictable and routine nature.

For the scientific probe there will be other workstations that enable the scientist to monitor the results being presented by a probe. Indeed, the scientist as an ultimate user of an aerospace system could undertake some of the duties of a spacecraft controller. For example, the scientist could initiate telecommands to operate the relevant probe, provided that no disturbing influences are experienced by other on board subsystems and experiments. Such an arrangement would be influenced by how and where the executive expertise is delegated and located. This can depend greatly upon the purpose of the spacecraft mission and the TTC standards which are employed.

For the pre-launch activities, the executive control of scientific and applications spacecraft can be maintained from the prime OCOE workstation by the senior engineer who is the test conductor. The senior engineer effects control from the OCOE test conductor console which has the master executive capability for system control during pre-launch operations.

For the pre-launch construction phases of the spacecraft, other test engineers can be engaged at OCOE observer consoles. These are workstations which do not have full executive capabilities. However, these workstations can be delegated control from the test conductor console. Therefore, all engineers engaged in spacecraft pre-launch operations can have a controlled access to the executive control interface.

The effects of telecommand transmissions are monitored by the space segment and transmitted to the ground segment as housekeeping telemetry parameters. These parameters enable the ground segment to monitor the telecommand actions from the housekeeping telemetry. Therefore, telemetry and telecommand facilities can function in a closed-loop mode of operation for safe overall system control.

The presentation of spacecraft housekeeping data at control workstation interfaces, if made in a user-friendly manner, will most likely be in engineering terms, rather than necessitating an analysis of binary data to be undertaken. Spacecraft housekeeping telemetry data can contain many parameters (hundreds). These are processed and the results are made available at the executive spacecraft control workstations. Such parameters are allocated an identifier which most often is in alphanumeric form. Thus, the same identifier should be adopted for pre- and post-launch activities. Acronyms of parameters associated with telemetry and telecommands are a case in point. Derived parameters which can be produced from the processing of telemetry data also require attention. A simple example of such a derived parameter could be watts ($W=VI$), with the telemetered parameters being voltage (V) and current (I).

Operation of an overall system is most efficiently implemented with active closed-loop control. That is to say, with the provision of facilities which enable the results of telecommand transmissions to be monitored with telemetry signals in a real-time environment. Linking these actions and allowing subsequent operations to be performed in an automated manner has benefits which will be detailed in Chapter 5. To enable the closed-loop control to be performed safely and efficiently, computers and associated software are employed to provide executive control facilities at workstations for spacecraft controllers and users.

4.6 EXECUTIVE CONTROL CONFIGURATIONS

The executive control facilities should enable the health of the space segment to be monitored and controlled from a central location. The post-launch ground segment will have to deal with flight dynamics tasks which embrace in-flight control activities. The pre-launch ground segment does not have to deal with the flight dynamics tasks, but, as described in Chapter 3, functional tests regarding the tracking capabilities are necessary and are aligned to spacecraft operations. In general, for both the pre- and post-launch ground segments, executive control is fundamentally a hybrid.

4.6.1 Pre-launch configuration

The pre-launch spacecraft system level testing includes the use of environmental test facilities. A thermal vacuum chamber is such a test facility which is normally operated under computer control. The testing of a spacecraft in such an environment can be considered to be performed with the executive control being provided by a hybrid configuration. This configuration is given in Fig. 4.9. The control computer of the thermal vacuum chamber controls the simulation of the space environment. The OCOE master test processor (MTP) controls the spacecraft via the OCOE telemetry and telecommand subsystems. The OCOE MTP also controls SCOEs which complement the test chamber in the simulation of post-launch environments. Liaisons between the workstations of the MTP and the test chamber are necessary to enable coherent tests to be performed.

4.6.2 Post-launch configuration

The post-launch executive can also be considered as conforming to a fundamental hybrid configuration. This configuration is given in Fig. 4.10. The spacecraft is controlled by the spacecraft control computer via the TTC stations. The flight dynamics computers deal with orbit determination and predictions, and provide the necessary orders to the spacecraft control computer for the attitude and orbital control of the spacecraft. This task can be performed with man–machine interface liaisons between the computer workstations. For this case, the executive hybrid is termed the operations and control centre (OCC).

4.6.3 Configuration comparisons

Commonalities between the hybrid configurations for the executive control of both the pre- and the post-launch Earth segment can be identified from Figs 4.9 and 4.10.

The requirements of both the OCC spacecraft control computer and the OCOE MTP to perform spacecraft control operations can now be confirmed as being

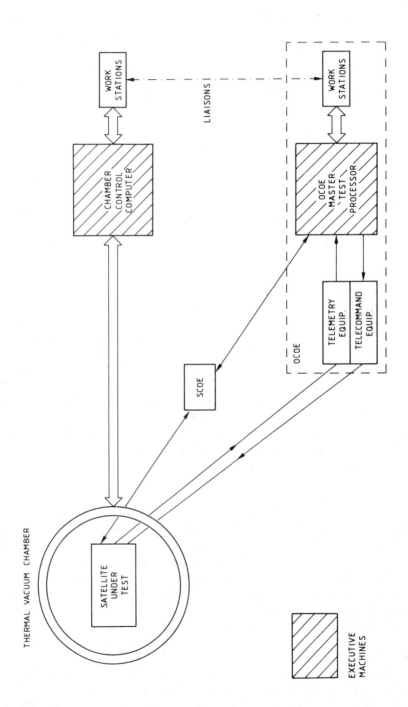

Fig. 4.9 — Pre-launch control hybrid configuration. The operation and test of a spacecraft in a simulated space environment is effected by hybrid control.

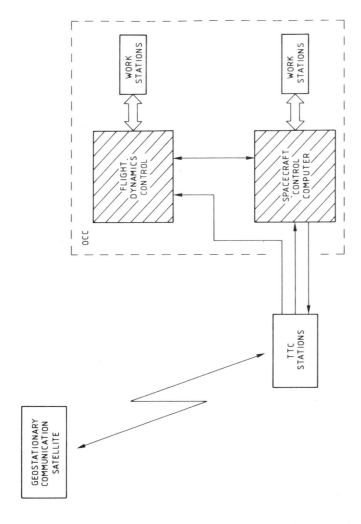

Fig. 4.10 — Post-launch control hybrid configuration. The hybrid control is undertaken from the operations and control centre (OCC).

similar, if not identical. Variations can occur if pre-launch operations during AIT are not a close simulation of the expected post-launch activities and the non-terrestrial environment. Furthermore, if the spacecraft functions satisfactorily after launch then any built in redundancies and failure recovery modes may be unnecessary for post-launch situations. Harmonizing the application and development of pre-and post-launch ground segments can certainly be considered as a comprehensive systematic approach exercise for spacecraft control.

For space exploration, such an approach, to some extent, is more difficult to implement. The simulation of an orbital path on a deep space probe can be more difficult, as can the simulated operation of the associated scientific probing instruments. This is natural when considering that if all was known, the mission would not be within the exploration category.

For the subject communication system, pre-launch operational control of the satellite is exercised by the OCOE, and after launch by the OCC. Whether the satellite is controlled by the OCC or the OCOE, it should respond in the same manner. The telecommands (control messages) which are transmitted to the satellite, and the responses encoded and transmitted as housekeeping telemetry from the satellite, have the same format in both pre- and post-launch cases. It should not be unrealistic, therefore, to utilize the same ground segment components of hardware and software to provide almost identical executive control interfaces.

5

System executive:
man and machine

The executive facilities required for spacecraft control are those which exercise all facets of operations with safety and efficiency, enabling ultimate user requirements to be met. Executive facilities are provided by machines and their human operators. The term executive machine relates to a computer system, and the term controller refers to their operators.

The executive requirement will be identical in principle for both pre- and post-launch operations. The executive facilities of such ground segments will need a man–machine interface (MMI) to be established and employed. Computer software will be a prime component of this control interface, providing facilities in the form of a user language. Mission control for a communication satellite is primarily an engineering task which is performed by post-launch satellite controllers. The pre-launch satellite controller (AIT test conductor) performs activities to test the ability of the satellite to meet the post-launch requirements. Both types of controller task are made easier if the dialogue and data presentation at workstations is accomplished by using engineering terminology.

It is worthwhile noting that a spacecraft engineer uses a workstation as a tool. The main task is naturally concerned with the functioning of on board subsystems. Interests in the complexity of the hardware and software which are being used to accomplish this task is rather academic to such a user. Thus, the flow of data through the layers depicted on Fig 5.1, resulting from perhaps a simple keyboard command, will not be readily apparent. However, the ease of accessing satellite data facilitated by suitable software construction will be very much appreciated by the user.

5.1 OPERATIONAL SOFTWARE

The operational software which is utilized by the executive machine can be composed of several layers, as shown in Fig. 5.1.

The highest level provides the man–machine interface (MMI) for engineers at workstations. The lowest level is the machine code interface which facilitates

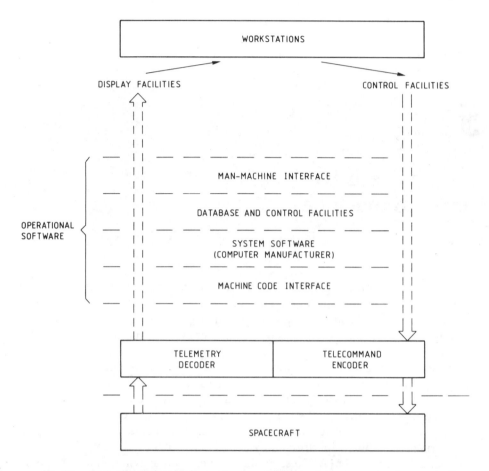

Fig. 5.1 — Basis of operational software structure. The layered structure makes the complexity of operations transparent to the users.

connections to the telemetry and telecommand baseband equipments. Ideally, the operational software should enable the workstation interface to be made by utilizing basic terms and symbols so that the control of the spacecraft can be efficiently handled. This is a prime function of the operational software. As previously mentioned, a dialogue needs to be made by employing engineering terminology to provide a user-friendly interface for spacecraft engineers. Furthermore, the interface, besides allowing engineering terms to be used, should also enable operations to be performed without the necessity of learning a complex computer language. As can be seen from Fig. 5.1, the computer manufacturers' utilities (system software) form an integral part of the operational software. In addition a database is required which defines the optimum expected on board subsystem conditions. These are compared with received and processed telemetry data.

In general, the greater the number of software layers, the more user-friendly the man–machine interface becomes. This allows a more usable result to be presented, but with the detailed intricacies of operation being transparent to the user.

5.1.1   User database

The database contains the information which is required by the operational software to perform the real-time operations. It consists of data in the form of tables and sets of statements in a high-level language, allowing automation of operations. This database is used by the operational software whilst control activities are in progress.

5.1.2   User facilities (control and display)

These facilities enable the executive machine to display the results achieved from the data processing functions of the operational software. They also enable control instructions to be relayed to the operational software by users.

5.2   PRE-LAUNCH EXECUTIVE

During the pre-launch phases of spacecraft operations, the overall checkout equipment (OCOE) master test processor (MTP) has executive control of all operations. The MTP controls the telecommand transmissions, the processing of housekeeping telemetry, and the distribution of data to the OCOE workstations. The MTP should have executive control of the special checkout equipment (SCOE). The SCOE workstations may be delegated control of a spacecraft's on board subsystems and its testing during some phases of spacecraft construction, but the executive control should be retained by the MTP. The executive user (the test conductor) who is located at a workstation (test conductor console) has control of the system operations. This engineer will employ the facilities that are provided by the operational software.

The following descriptions of the manner in which spacecraft control and monitoring are fulfilled is representative of all phases of a spacecraft's life-cycle. The descriptions of the executive control facilities use the pre-launch activities and OCOE as a model. It should be emphasized that the operations and control centre (OCC) and the telemetry, tracking and control (TTC) stations utilized during the post-launch activities perform many functions which are identical to or very similar to those of the OCOE.

5.3   EUROPEAN TEST AND OPERATIONS LANGUAGE

The European test and operations language (ETOL) will now be described to demonstrate how a database and user control facilities can be provided and, specifically, the processing functions which are required to enable system control to be maintained.

Most of the ETOL software was developed for use with the OCOE employed during spacecraft construction. This software has been used by satellite integration engineers during the construction of many European satellites.

ETOL provides facilities for constructing and operating the database that is required to test and verify spacecraft operation. The major elements of this ETOL software are now described in a general form. Should readers require additional information, reference should be made to the *Appendix* and the *Bibliography*.

5.3.1 ETOL facilities

ETOL comprises on-line and off-line facilities. The on-line facilities are those software modules engaged in the operational processing cycle. The on-line facilities process the incoming data in real-time. This, as will be explained, is a finite time later than events occur on board the satellite. The off-line facilities are programs which enable the construction of the database used by the on-line facilities.

ETOL enables the following operations to be undertaken:
(a) Monitoring the operations of:
   - Spacecraft
   - EGSE elements.
(b) Generation of orders:
   - Spacecraft telecommands
   - Instructions for EGSE elements.
(c) Automatic testing and control of:
   - Spacecraft
   - EGSE elements.
(d) Distribution of the results from (a), (b) and (c) to workstations for display purposes.
(e) Executive control function for (a), (b), (c) and (d) at workstations.

5.3.2 Off-line features

Before the executive control facilities can be operated, the database needs to be constructed. This is achieved by an ETOL user without the direct need for software specialists. ETOL provides three facilities which enable the complete database to be constructed off-line, without disturbing on-line operations. These are:

MTGP	Monitor table generation program. Produces the database associated with telemetry and telecommands.
SEQUENCE COMPILER	Compiles ETOL statements to produce programs for automated test procedures. These are termed ETOL test sequences.
PICTGEN	Colour synoptic PICTure GENeration program. Produces the associated database for information displays.

5.3.3 On-line features

The following ETOL software modules will operate in real-time, that is to say as soon as the relevant data become available. These operations are dependent upon reception of satellite telemetry data.

MONITOR	Enables the received telemetry data to be compared with the expected values contained in the monitor tables. The results of this process, primarily the unexpected conditions, are relayed to the output devices.

SEQUENCER   Enables the ETOL test sequences to be executed
                             (started and operated).

TVPICT   Updates the colour synoptic pictures in conjunc-
                             tion with the operation of MONITOR and the
                             ETOL test sequences.

5.3.4 ETOL architecture overview

An overview of the ETOL architecture is depicted in Fig. 5.2. From this figure it will
be seen that a software module ENVIRO is used to transfer the database generated
off-line to the on-line environment, thus enabling real-time monitoring and control
to take place. Control and display devices are employed for both on- and off-line
operations, and are interfaced to the operational software by the modules shown in
Fig. 5.2.

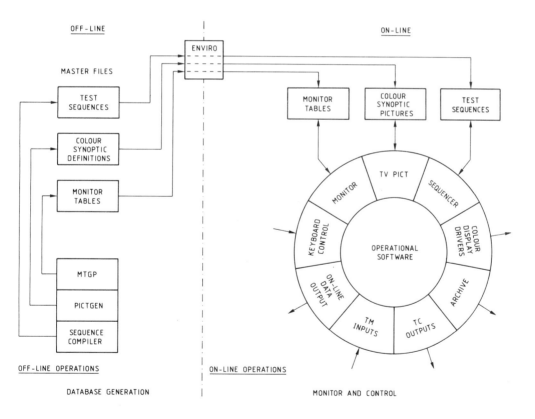

Fig. 5.2 — ETOL architecture overview. Off-line and on-line software modules that enable
database generation and operation.

5.4   ETOL DATABASE

Utilization of the user facilities enables the following database elements to be constructed. The descriptions which follow are specifically aligned to spacecraft telemetry and telecommands, but such database elements can be utilized for the operation and control of the SCOE.

The ETOL database elements are normally constructed off-line. Modifications to the monitor table database elements can be performed on-line whilst spacecraft operations are in progress.

5.4.1   ETOL monitor tables

ETOL monitor tables are part of the database which is assembled to enable spacecraft operation and performance to be controlled. Monitor tables are constructed so that received spacecraft telemetry parameters can be compared with expected values and conditions. Telecommands transmitted to the spacecraft are also defined in a monitor table. A telecommand transmission is recorded for subsequent evaluation of a related telemetry parameter; an expected result.

Telemetry monitor tables

The telemetry monitor tables are subdivided into sections, each of which can represent an on board subsystem. Subsystem parameters that are transmitted as telemetry values are assigned an alphanumeric code and label in the tables. It is this identification that allows the grouping of telemetry parameters into sections which are relevant to on board subsystems. These monitor tables define the conversion of received telemetry parameters into engineering values. The tables also define the operational limits which must be applied to these engineering values. When a received telemetry parameter exceeds a defined operational limit or condition, an output is made giving its value or status, together with what was exceeded.

Telemetry parameters fall into two basic data categories:

Analogue parameters

The source of these parameters will be on board sensors that have sampled analogue values which, for example, may represent voltages, currents and temperatures. The analogue values will be converted by the on board spacecraft subsystems from engineering values into telemetry data. (As explained in Chapter 4, the telemetry data are encoded into telemetry words/bytes which are transmitted within the telemetry format.) After reception and decommutation, the ETOL processing functions will reconvert the telemetry data into engineering values.

For the reconversion of telemetry data into engineering units, a conversion curve is employed. The curve also performs the function of calibration. The ETOL calibration curve performs this dual function and is part of the telemetry monitoring table definition. The conversion/calibration process is shown in graphical form in Fig. 5.3 and explained as follows:

An on board temperature sensor gives a voltage output, thus the sensor has a calibration curve of temperature against voltage. The voltage is converted by the on board telemetry encoder to give a telemetry value which is inserted into the telemetry format. Upon reception and decommutation, this value is then reconverted back into a temperature value by the ETOL software processing, using the

SPACE SEGMENT
(TEMPERATURE SENSOR)

EARTH SEGMENT
(ETOL PROCESSING)

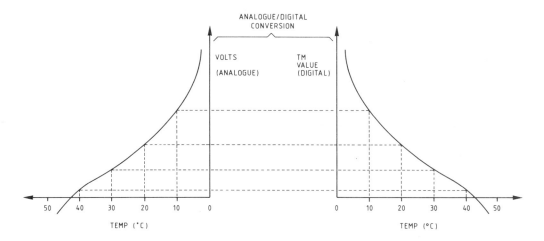

Fig. 5.3 — Telemetry conversion/calibration curves. An analogue parameter sensor produces an output voltage which is converted to a telemetry value (digital). A corresponding ETOL curve is used to reconvert the telemetry value to an analogue value. *Note*: To aid explanation *x–y* axes of the curve on the right hand side are not shown according to convention.

calibration curve. The ETOL calibration curve is a series of co-ordinates of several points along the curve which are held in the monitor table. Because the number of points that can be held is limited, some linear interpolation will be performed when the conversion from digital-to-analogue values takes place as part of the operational software processing.

Each analogue parameter may have limits attached to it. These can be upper and lower limits which are compared against a nominal value. In addition, analogue parameters can have danger limits attached to them which, if exceeded, results in automatic corrective or safety actions being implemented. Another limit that each analogue parameter can have associated with it is termed a delta. This ETOL delta is a change of value between consecutive telemetry samples relating to a particular monitor parameter. If the defined delta limit is exceeded, an output message results.

An example of a calibration curve with associated limits is shown in Fig. 5.4. The telemetry value in the range 0–255 relates to a single telemetry word of eight bits. The coordinate points employed, demonstrate how linear interpolation is achieved.

Digital parameters

These parameters result from the sampling of binary status registers of the on board subsystems. These registers contain data that is relevant to on board processors and the status of on board systems. In the most simple case, a telemetry channel can be used for monitoring the ON/OFF status of an on board unit. The operational

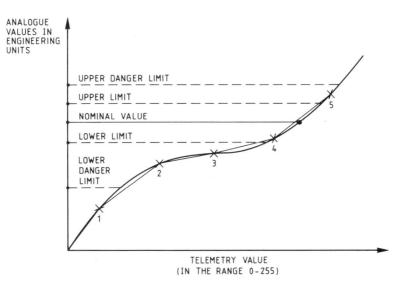

Fig. 5.4 — ETOL calibration curve. Analogue parameter calibration curve, having five
coordinate points, with the application of ETOL limits about a nominal value.

software processing can convert the binary logical value of '1' or '0' into the
engineering status ON or OFF respectively.

Telecommand monitor tables
The telecommand monitor tables define an alphanumeric code for each telecommand action and assign this definition to the associated telecommand frame. In
addition, the expected conditions for a successful telecommand execution are also
defined to allow this to be monitored via the spacecraft telemetry data. The entities in
the telecommand monitor tables will allow cause and effect comparisons to be made
by the operational software.

The telecommand and telemetry monitor tables operate in unison. This unified
operation facilitates the closed-loop control of the spacecraft by the telemetry/
telecommand interfaces. Examples of both telecommand and telemetry monitor
tables are given in the *Appendix*.

5.4.2   ETOL colour synoptic pictures
The ETOL database enables spacecraft telemetry data to be displayed in colour
synoptic picture form. Telemetry parameters are displayed in these pictures which
may be constructed in the form of circuit diagrams, block diagrams, or in alphanumeric form only. The pictures can also be constructed to be a combination of these
types of display. The colour diagrams or pictures represent spacecraft configurations
and conditions. The pictures are updated dynamically by the software operating on
telemetry data and thus represent the current spacecraft status in a synoptic pictorial
form. There are a number of symbols available for representation of a picture, for

example switches, boxes, text, vertical and horizontal lines. Values and status of telemetry parameters can be used to change switch settings, illuminate lamps, and change colours. A library of colour synoptic diagrams can be constructed. The library can contain pictures representing overall spacecraft status and operations. Other pictures can then expand areas of this picture, providing a zoom-in effect. Examples of colour synoptic pictures which demonstrate the zoom capability are given in the *Appendix*.

Colour synoptic pictures can also be constructed to enable information which is received from the SCOE to be similarly displayed. Furthermore, pictures which are not a display of received data can also be constructed. Examples of these are also given in the *Appendix*. ETOL colour synoptic pictures are particularly powerful elements of a user friendly interface which complement other display facilities.

5.4.3   ETOL test sequences

The term which is given to the ETOL facility that enables automatic testing and control of the spacecraft to be performed is a test sequence.

An ETOL test sequence is a set of instructions/statements which can be assembled to meet specific requirements. These statements provide users with facilities that are additional to the ETOL functions inherent in monitor tables and colour synoptic pictures. Test sequences are written to function in unison with and complement these other ETOL features. Several test sequences can operate together, allowing spacecraft tests and operations to be performed in a coherent manner.

To enable the scope of the ETOL test sequence statements to be appreciated, the following examples of the functions which can be performed are given:

- Send telecommands to switch the power of an on board subsystem ON or OFF.
- Send telecommands to an on board subsystem to perform various measurement functions.
- Send commands to a SCOE to stimulate sensors.
- Verify expected conditions.
- WAIT statements to allow actions to be timed.
- Perform simple arithmetic operations.
- CONDITIONAL statements.
- Display the results of test sequences to selected output devices.

ETOL test sequences can initiate telecommands and monitor the telemetered results. These actions, together with other statements, allow automated tests and operations to be performed. It has been stated that monitor tables enable spacecraft telecommand actions to be monitored via telemetry. Telecommand parameters have expected status values linked to them. Thus, if a telecommand is sent, this information is retained, and when telemetry processing is performed, the parameter, possibly a status bit, will be compared to the expected value defined in the monitor table. Several of these actions can be linked together by a test sequence, making command initiation and transmissions automatic, and enabling the result processing to be automated also. If the results are correct, the test sequence continues; if the results are incorrect, then reports are made on display devices to draw the attention of the user, and the test sequence is suspended pending a manual intervention.

Additional information regarding ETOL test sequences is given in the *Appendix*.

5.5 ETOL CONTROL AND DISPLAY FACILITIES

The ETOL control and display facilities provide the closed-loop man–machine interfaces at workstations for spacecraft operations.

During spacecraft operations, the facilities display the appropriate results on output devices. These output devices can be a mixture of visual display units (VDUs), colour monitor screens, printers, and plotters. The devices used depend upon the configuration of the executive machine.

The operation of the monitoring tables automatically results in the output of anomalous conditions and values. If results are as expected there is no output. Such an operation is especially effective during final testing activities. However, there are occasions when a value is required to be registered even if it is within limits. ETOL also enables this to be performed. It is a facility that a user can instigate, and this is relevant to monitoring control.

ETOL keyboard commands that are entered at workstations, provide users with facilities that operate on-line. They fall into specific categories according to function, and are:

- ETOL monitoring control. These apply to the operation of the monitor table database. Monitor control commands allow some updating of this database during its operation.
- ETOL telecommand control. These commands enable set-ups to be made and controlled, allowing telecommand transmissions to be undertaken.
- ETOL test sequence control. These apply to test sequence operations, including start and stop commands.
- ETOL output device control. These commands enable utilization of the devices required, and which parameters should be displayed.
- ETOL System Control. These commands address facilities that are system oriented.

Further information regarding these ETOL facilities are given in the *Appendix*.

5.6 ETOL OPERATIONS

To enable the ETOL functions to be performed in a coherent manner requires that the operational software performs methodically. Some ETOL functions will be repetitive, with the reception and processing of spacecraft telemetry data being the prime example. ETOL enables the repetitive functions to be performed in a synchronous manner. This method of operation contributes to a user-friendly man–machine interface.

A main factor relating to spacecraft telemetry is the PCM bit rate of the video signal. This, or a derivative, can be the driver of operational processing cycles. The regular analysis of received telemetry data will be of great importance for safe and efficient overall system control. For spacecraft that comply with the ESA PCM telemetry standard and employ ETOL for executive control, the operational processing cycle runs at the telemetry format frequency. This operational processing cycle is often termed the checkout cycle.

5.6.1 Data buffers

The telemetry data transmitted from a spacecraft are in the form of a continuous serial stream. The operational software being executed in the executive computer has to receive these data and process them as quickly as possible. This is achieved by utilizing two input buffers with a software 'switch' determining which buffer receives the data at any given time. These buffers temporarily store the raw (unprocessed) telemetry data on a format-by-format basis. Because the data are received in the form of a continuous stream, as soon as the software switch to a buffer is activated the data in that buffer will be overwritten by fresh data. A full input buffer generates an interrupt to the processors to indicate that a complete telemetry format is present and ready to be processed. A third buffer (processed data buffer) is used to store the processed results of the engineering values produced from the telemetry data. The values held in this buffer may then be subject to further processing. The operation of these buffers is explained with the aid of Fig. 5.5 as follows.

At time zero, the spacecraft starts transmitting format 'n' which is loaded and stored in raw data buffer No. 1. The following format '$n+1$' will be loaded and stored in raw data buffer No. 2. When raw data buffer No. 1 is full, that is at the end of format 'n', the data will be converted from telemetry data to engineering values and stored in the processed data buffer. Raw data buffer No. 1 will then be available to receive and store the data of the next format '$n+2$' when it is transmitted. When raw data buffer No. 2 is full at the end of format '$n+1$', the conversion process will be activated again and the processed data from format '$n+1$' are stored in the processed data buffer. The data held in the processed data buffer are available to the operating software for the functions of, for example, limit checking and output to displays. This sequence will be repeated for successive formats whilst the spacecraft continues to transmit data. This process which utilizes two raw data buffers in the manner described is termed double buffering.

5.6.2 Operational processing cycle

Besides the buffering process described, three major functions associated with the database are executed during the operational processing cycles. These functions, as previously defined, are the monitoring tasks, test sequence operation and colour synoptic picture updating, effected by the ETOL modules MONITOR, SEQUENCER and TVPICT respectively.

These major functions are performed during the operational processing cycle in a period that can vary. The period will depend on how many monitor parameters, conditions, and status information need processing and reporting on an output device which, in turn, is dependent on the number of:

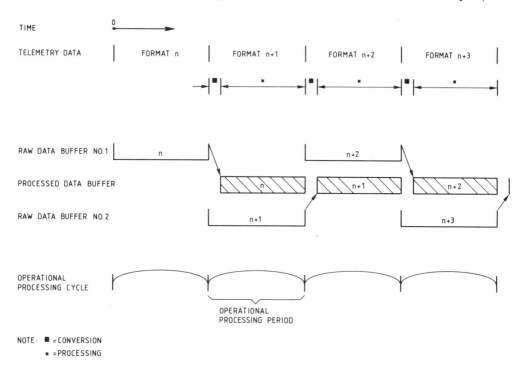

Fig. 5.5 — Buffer operations. Correlation of buffer activities dealing with telemetry data during the operational processing cycle.

- telemetry parameters that are outside the expected limits;
- telecommands that do not exercise on board functions as expected;
- test sequences being operated; and
- colour synoptic pictures that require updating.

The major activities that are performed during an operational processing cycle of the software used on the ESA European communication satellite (ECS) OCOEs are given conceptually in Fig. 5.6.

Other tasks may also be performed during operational processing cycles. These can be classified as special processing tasks and background tasks. Special processing tasks are those which are required to perform non-standard and special tasks that cannot be performed by ETOL alone. In such a case, a special processor routine would be written to operate in an intermediate layer of the software. An ETOL statement would call the routine which, upon completion, would return its result to the ETOL processing facilities. Special routines are normally executed during test sequence operations and are therefore part of the major processing functions.

Background tasks are performed in the remaining period of a checkout cycle, after the other tasks have been completed. The background tasks can encompass the compilation and writing of test sequences and the updating of monitor tables, such as changing of limits.

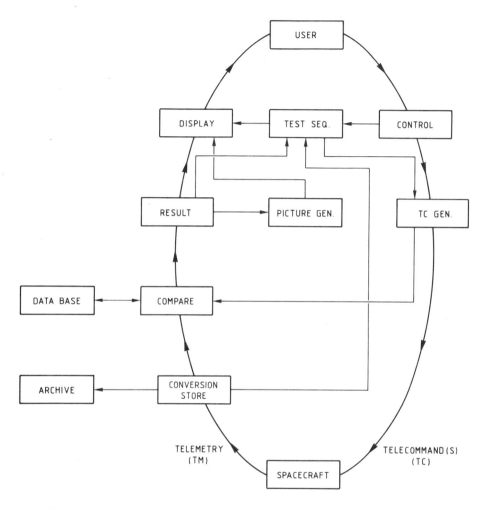

Fig. 5.6 — ETOL processing cycle — functional concepts. Major software events that are
undertaken during spacecraft operation.

The processes which have been explained take place during operational processing cycles. That is when the raw data buffers are being filled and the contents of the
real-time processed data buffer are being used. An overview of these operations is
shown in Fig. 5.7.

5.6.3 Event timing — ETOL application

An event which happens in real-time in the strictest definition is the time at which the
event occurs. However, from operational aerospace system aspects, with the post-
launch space and ground segment operations in particular, this interpretation is not
necessarily the case for the executive control workstation located within the post-
launch ground segment. An event which is considered to happen in real-time within a

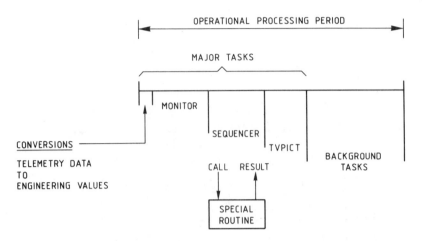

Fig. 5.7 — ETOL processing tasks. During an operational processing period software tasks are performed in the defined order which is portrayed.

ground segment can be somewhat historical from a space segment standpoint. It does, of course, take time for telemetry data to be received at the ground segment after transmission from the space segment, and this is termed propagation time. For a geostationary satellite the propagation time is short; for deep-space exploration probes it can be significant. In fact, for the ground segment, no space segment activities are monitored or initiated in compliance with the exact definition of real-time. The ETOL processing is performed as soon as the required data become available in a suitable form, that is when a raw data buffer contains a complete telemetry format.

The operational processing cycle explained has been utilized extensively for the control of European spacecraft during their construction. Satellites produced within the ESA ECS programme have TTC subsystems which operate with the following telemetry characteristics:

Telemetry bit rate — 160 bits per second
Telemetry word   — 8 bits
Telemetry frame   — 32 words
Telemetry format   — 12 frames

Therefore, the duration of an ECS telemetry format is:

$$\frac{\text{Total number of data bits/format}}{\text{Telemetry bit rate}} = \frac{8 \times 32 \times 12}{160} = 19.2 \text{ seconds}$$

It is this duration of 19.2 seconds that is the operational processing cycle period for ECS.

Therefore, by using ETOL to monitor and process the activities of ECS models, fully processed data can be available with a delay of between 19.2 and 38.4 seconds. The period is dependent upon the telemetry sampling and the processing used. For example, referring again to Fig. 5.5, an event could occur early in format 'n' but the data may not be available until the end of the operational processing period. These conditions would give a delay of approximately 38 seconds.

The user at an ETOL workstation sees an event some time after it has happened. This delay is due to a combination of several factors:

- Telemetry sampling delays on board the space segment.
- Propagation time of the telemetry signal.
- Buffering of telemetry data in the raw data buffers.
- Processing of data during process periods.

For ECS, calculations will reveal that the propagation time for telemetry data is in the order of 0.13 seconds. This period can be analogous to the pre-launch condition, depending upon such factors as the geographical position of the ground segment. Therefore, the following description can be applicable to both pre- and post-launch satellite operations.

5.6.4   Event synchronizations

The time factors which have just been described demonstrate that control of the space segment should be considered from total system concepts during the design and definition of aerospace systems. Closed-loop control should be dominated by synchronizing events with actions if safe and efficient system control is to be achieved. The longest closed-loop will be when a user makes observations, considers the results, and then undertakes actions. This action is generally termed manual control, and can be necessary for some space segment operations. If the executive operations allow events to be synchronized automatically, then this can reduce response times to monitored events.

The control of the space segment by synchronizing telecommand transmissions and verifications by the ground segment will now be considered. The control of ground segment elements can deserve similar attention if systems are to be controlled in a coherent manner from an executive centre.

The transmission of a telecommand is made to instigate some event either directly or indirectly. Direct actions are usually associated with the configuration of on board spacecraft subsystems. Indirect actions usually result in processing that can be undertaken by on board spacecraft computers. For example, direct telecommand actions can perform switching functions, whereas indirect actions can constitute the loading of software for an on board processor. The results of the telecommand transmissions can be verified by more than one method. Command actions, or executions as they are often termed, can therefore be complex from a verification standpoint.

Taking a relatively simple example of a telecommand that switches an on board unit ON and OFF, the direct telemetry channel associated with this command will most likely be a single binary bit which can change from a '0' to a '1' status, and vice versa. Such an action can also be monitored indirectly by another telemetry channel that records a change in the current of the power supply bus. Furthermore, the encoded command within a telecommand frame (ESA PCM telecommand standard) can also be monitored by housekeeping telemetry. This enables a verification to be made that the telecommand transmission was error free. Now referring to Fig. 5.8,

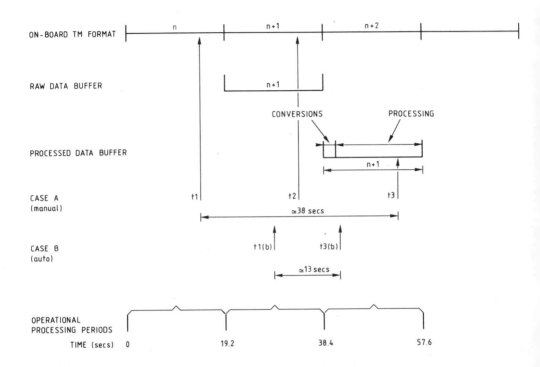

Fig. 5.8 — Event synchronization with telemetry format. Comparison of manual and auto control for efficient satellite operations.

the results of the telecommand transmissions are analyzed in some detail. The periods which are relevant to telecommand verifications are approximate and indicative to demonstrate time differences between manual and auto control. The examples presented are applicable to satellite construction (ECS models), so the propagation times between telecommand transmission and reception are not relevant.

Case A: Manual control

Table 5.1 shows the timetable of events concerning the transmission of a telecommand and its subsequent processing.

For this case, the period t_1 to t_3 would be approximately 38 seconds. After time t_3, manual actions could be undertaken subsequent to the analysis of the results by an engineer.

Table 5.1 — Manual control event timing

Time	Event
t_1	Telecommand transmission/reception/execution.
t_2	Sampling of the telemetry channel which is directly associated with the telecommand that was transmitted at t_1. This sampling is a function of the on board TTC subsystem encoder operation.
t_3	Result available on executive controller output devices after the major processing functions have been completed.

If immediate subsequent action was necessary after telecommand transmission and verification then it could not have been undertaken before t_3 with manual verification. These events could be more efficiently implemented by the judicious timing of events. If the telecommand transmission/reception/execution is earlier than the time when the relevant telemetry sample is taken, then the data will be available in the telemetry format being constructed. Therefore, if the telecommand transmission can be synchronized to occur a short time before the telemetry sample, the elapsed time before verification can be reduced. Furthermore, if the actions emanating from the verification processes can be implemented almost immediately afterwards, another reduction in the delay between telecommand transmission and resulting actions can be achieved. The next example, again refer to Fig. 5.8, demonstrates how such events can be made to occur.

Case B: Auto-control
Table 5.2 shows the significant event times concerning a telecommand transmission and subsequent processing.

Table 5.2 — Auto control event timing

Time	Event
$t_1(b)$	Telecommand transmission/reception/execution synchronized with centre of telemetry format
t_2	As case (A)
$t_3(b)$	Result available in processed data buffer for automatic action by major processing facilities.

In this case, time period $t_1(b)$ to $t_3(b)$ would be approximately 13 seconds. After time $t_3(b)$, subsequent actions can be undertaken without significant delays if automated as part of the major processing activities.

ETOL provides facilities that enable telecommand transmissions and verifications to be made complying with the Case B example. This allows spacecraft test and operations to be safely and efficiently performed by using ETOL and minimizing manual activities.

5.7 WORKSTATION COMMONALITIES

The approach of applying the systems engineering techniques to optimize working practices will be of importance at the man–machine interface. This will be most prominent at the executive control workstations of the OCOE and OCC. In fact, the facilities located within the executive machines do require careful consideration when designing an aerospace system. This is especially true if efficient, safe, and cost effective solutions are to be employed. So with the end-to-end approach of the systems engineering technique, the facilities which are presented by an aerospace system executive are extremely important and do necessitate rationalization to give an efficient overall programme.

The ETOL operational processing cycle, which encompasses the operation of the layered software, is driven by the spacecraft telemetry timing. Therefore, if the spacecraft telemetry format is exactly the same for pre- and post-launch operations, which is normally the case, the executive software processing cycle can be identical. This gives the possibility for the operational software to be the same for pre- and post-launch executive controllers. Certainly, similarity of nomenclature applied to telemetry parameters and telecommand orders and instructions should be achieved, even if the executive controller system level architectures are different for pre- and post-launch operations.

As a systematic approach is being presented, it was considered that ETOL was a good medium to express the methodology for efficient spacecraft control. Telecommand and telemetry operations are normally identical for geostationary communication satellites for both pre- and post-launch activities. ETOL has been taken as an example because this language is used extensively for pre-launch operations, and for some post-launch activities, as will be described in Chapter 6. ETOL has also been utilized during the construction of scientific satellites engaged upon probing and experimental missions.

The housekeeping tasks that are necessary to keep a spacecraft operating as required are basically identical for both scientific exploration and the commercial exploitation of space. Thus, opportunities exist for the rationalization of techniques and working practices that are applied to spacecraft control.

6

System rationalization: development appraisal

Research into and the application of technologies associated with the non-terrestrial environment has produced some duplications. This can be regarded as a natural consequence of a variety of early development activities. Duplications can occur between different spacecraft programmes and within space exploration and exploitation endeavours. The duplications need to be reviewed so that similarities can be identified, enabling them to be resolved in an efficient manner. To exploit technical similarities enabling practical utilization to be achieved needs a rationalization process to be undertaken. This process, with a comprehensive and systematic approach, can produce technical standardization, enhancing opportunities for a reduction in the amount of development duplication. The implementation of technical standards will lead to common working practices which, if correctly implemented, give safe and efficient spacecraft control by the judicious re-use of technical developments. This chapter reviews some practicalities and then presents areas where rationalization deserves consideration for satellite design and control.

6.1 PRE- AND POST-LAUNCH COMPARISON

The rationalization process applied to pre- and post-launch activities should obviate the need for some duplications. The architecture of a communication system spacecraft does not change after launch; it is identical for both pre- and post-launch operations. In principle, the architecture of the ground segment also remains unchanged, but can vary in composition for pre- and post-launch situations. The standard concept which is maintained is that the space segment is always under the executive control of a ground segment. This is particularly relevant to the monitoring and control of a communication satellite by telemetry and telecommand.

A rational approach to the operation of the space and Earth segments of an aerospace system should bring benefits. From a systems engineering standpoint, requirements for system construction and operation occur during the following phases of a satellite's complete life cycle in the sequence given:

Phase 1. Commence with on board unit construction and test.
Phase 2. Continue through
     (a) Satellite AIT
     (b) Launch
     (c) Orbital operations
Phase 3. Terminate at the end of life in orbit, between 10 and 20 years after launch.

Health monitoring and control of the satellite will be necessary for all activities defined in phase 2.

Significant rationalization and standardization of the pre- and post-launch ground segments can be achieved. Commonalities between pre- and post-launch ground support equipment need to be defined and maximized. Such actions should enable a single and coordinated design and development task to be undertaken to establish a comprehensive use of ground segment elements. The application of telemetry and telecommand standards for pre- and post-launch operation is identical. The prime difference for the reception and transmission of telemetry and telecommand signals by the pre- and post-launch ground segments is that of the RF power required to accommodate the changes in distance between the relevant ground segments and the spacecraft. The transmission medium for telemetry and telecommand signals can, as explained in Chapter 4, differs to some extent for pre- and post-launch spacecraft operations. However, commonalities do exist for both situations and these can be exploited further if prompted by rationalization. Such actions could give an increase in the standardization of pre- and post-launch ground segments and their operations, enabling efficient and safe spacecraft control during its complete life-cycle to be achieved.

6.2   PRE-LAUNCH DEVELOPMENTS

Respecting operational timescales, the application of pre-launch ground segment technologies are now reviewed. The adoption of rationalizing techniques has been applied during the construction of the majority of European communication satellites. Certainly, during satellite construction and all pre-launch testing, a significant degree of standardization which permits comprehensive pre-launch satellite operation to be performed has been achieved. The EGSE used on European communication satellite programmes has been rationalized to a significant extent, and the overall checkout equipments (OCOEs) have functioned with the European test and operations language (ETOL). As stated in Chapter 5, this language allows satellite construction engineers to operate the satellite without this task being onerous to their testing activities; ETOL presents its users with a friendly man–machine interface. The language allows databases to be constructed so that during test operations, anomalous results can be highlighted, thus permitting testing to be performed on a GO/NOGO basis.

The ETOL database construction can be achieved without the need of specialized computer software expertise. An automated ETOL system provides for repeatability of testing and minimizes user intervention. In addition, ETOL allows automated testing methods to be developed during spacecraft construction, and standardized, and consequently could be re-used for post-launch activities. Furthermore, similar versions of the high-level user language (ETOL monitor tables, synoptics, and test sequences) have been re-used with minor modification on several of the European communication satellite programmes undertaken by various authorities. The following review again takes the ESA ECS programme as a baseline for explanations of how ground support facilities can be re-used. The explanations will demonstrate how EGSE items can be employed during the different phases of a satellite programme. In addition, it will be apparent that re-use of design and developments can be achieved across programmes.

6.2.1   Spacecraft unit and subsystem level testing

For the construction of the spacecraft, electrical ground support equipment is necessary at unit, subsystem, and system (spacecraft) levels of test. During spacecraft unit level testing, the equipment consists of general purpose laboratory instruments which are supplemented with special-to-test jigs and equipments. Following unit level testing, the basic EGSE architecture for subsystem level testing is as shown in Fig. 6.1. The architecture allows for the subsystem to be tested and

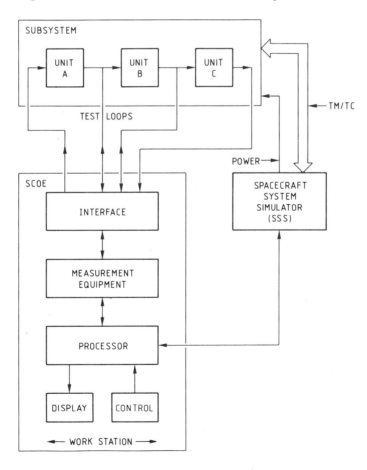

Fig. 6.1 — Basic EGSE architecture for spacecraft subsystem level testing.

verified whilst it is presented with an interface which is representative of that which it will experience when it is integrated into the spacecraft.

There are two prime components of this architecture; the spacecraft system simulator (SSS) and the specific checkout equipment (SCOE). The SSS provides the electrical power and the telemetry/telecommand interfaces to the subsystem under test. In fact the SSS in this case simulates the on board PSS and TTC subsystem. For the ECS example, the term service module simulator would be more accurate, but for a more general·approach the term SSS is employed. The SCOE is the controller of the on board spacecraft subsystem under test and the SSS. The SCOE is normally a computerized system and performs closed-loop test and performance measurements. During testing, an engineer at a SCOE workstation can configure the subsystem under test from the SCOE via the telecommand interface to the SSS. Correct operation of the telecommands is verified via the telemetry interface of the SSS. After the correct configuration has been confirmed, test and measurements can be performed by the SCOE utilizing the test loops indicated in Fig. 6.1. This architecture can form the basis of all SCOE designs for the independent (subsystem level) testing of on board subsystems.

6.2.2   Spacecraft system level testing (intelligent SCOEs)

The system level spacecraft EGSE that has been described in previous chapters is the pre-launch Earth segment of an aerospace system. These descriptions used the case of relatively simple designs. However, SCOE operations at spacecraft system level have taken place under the executive control of the master test processor (MTP); the computer facilities of the overall checkout equipment (OCOE). The OCOE MTP provides executive capabilities, enabling telemetry to be decoded, processed, and distributed, and telecommands to be encoded and transmitted. The OCOE allows telecommand actions to be monitored by analysis of the spacecraft housekeeping telemetry, giving closed-loop control. Furthermore, telecommand messages are analyzed by the OCOE during transmission, and if errors are detected, actions are taken to ensure that rejections occur. The man–machine interfaces for all these activities are made at the OCOE workstations.

The increasing complexity of on board spacecraft subsystems has resulted in the production of more intelligent SCOEs. In other words, most SCOEs are equipped with microcomputer facilities to cope with the consequences of more complex test requirements. As a result, the EGSE architecture must allow for the increased intelligence to be distributed and utilized in a coherent manner. If this criterion is adequately satisfied, then safe and efficient spacecraft control and operation can be performed. The architecture given in Fig. 6.2 provides for this requirement. The MTP retains the executive control for spacecraft operations, so the basic concepts remain unaltered. To enable ordered exchanges to take place between the MTP and other intelligent EGSE components, an EGSE interface protocol is required. A local area network (LAN) could be included in the system to provide the basic services for this traffic. The LAN would facilitate the connection of the MTP to the SCOEs, and the OCOE telemetry and telecommand subsystems. Furthermore, the EGSE user workstations, the test conductor and observer consoles could all be connected to the LAN. Should occasions occur when it is necessary for the SCOE to be remotely located from the OCOE, then the architecture which incorporates the LAN could be

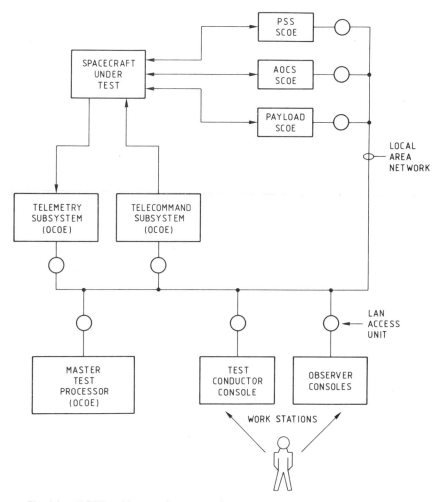

Fig. 6.2 — EGSE architecture for spacecraft system level testing. Intelligent SCOE.

adapted to meet any necessary requirement for remote operations. These could be met by interfacing the LAN to a network using a more advanced protocol such as X25 (packet techniques).

6.2.3 Spacecraft system level testing (unintelligent SCOEs)

To enable EGSE architecture to be implemented with relatively unintelligent SCOEs but taking advantage of a defined interfacing protocol, the configuration given in Fig. 6.3 can be adopted. In this case, a star configuration is employed. Again, the MTP retains executive control with most of the system intelligence residing within the MTP. The EGSE interfacing protocols could be a subset of those employed when the architecture includes a LAN. This configuration can be rather

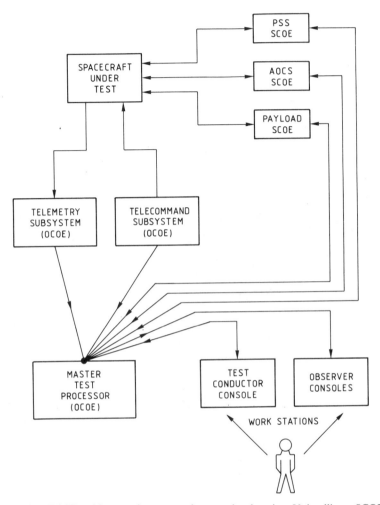

Fig. 6.3 — EGSE architecture for spacecraft system level testing. Unintelligent SCOE.

constricting if there are activities which require the SCOEs to be significantly remote from the OCOE.

6.3 EGSE ARCHITECTURE COMPARISONS

The EGSE architectures which have been described are based upon modular construction methodologies employing software layering techniques.

A comparison of the EGSE architectures described show how SCOEs can be used at both spacecraft subsystem and system levels of test. Designs can be made common and should, in most cases, be able to incorporate technologies developed during unit level testing. At the least it can be expected that the same type of commercially available laboratory instruments used at unit level testing are integrated into SCOEs.

The OCOE, SSS, and unit level TTC subsystem test equipments must have facilities for telecommand generation/transmission and telemetry reception/decommutation. This gives a commonality which with rationalization should enable design and developments to be undertaken so that EGSEs can be re-used for each pre-launch testing level (unit, subsystem and system) within a spacecraft construction programme.

The composition of the EGSE which can be established for these different levels of test are summarized as follows:

Unit level   : Laboratory instruments plus special test aids, jigs, and equipment.
Sub-system level : SCOE which incorporates unit level EGSE components, together with the SSS.
System level   : SCOE and OCOE

The re-use of EGSE designs across spacecraft programmes is possible, but is heavily dependent upon the standardization of on board designs between programmes. The on board TTC subsystems do have mandatory requirements to interface with post-launch ground segments and thus can exhibit standard interfaces for telemetry, tracking, and command functions. Therefore, there is an opportunity for having a standardized and re-usable pre-launch EGSE which is dominated by the telemetry and telecommand requirements. This can enable rationalization to establish common designs for:

* On-board spacecraft TTC subsystem test equipment (unit level).
* TTC equipments for the spacecraft system simulator (subsystem level).
* OCOE telemetry/telecommand subsystems (system level).

The commonalities in unit level, subsystem level, and system level EGSE items can embrace computer software besides equipment hardware. This should allow these EGSE components to be utilized across spacecraft programmes. Therefore, the re-use of EGSE designs applied equally to both hardware and computer software should enable full implementation of rationalization for spacecraft control for pre-launch purposes within and between programmes.

6.3.1 Interface considerations

If a systems engineering approach is applied then technical standards and common methodologies can be incorporated into Earth segment design and operation. Such an approach will be strongly dependent upon technical interfaces between system elements within both Earth and space segments.

The following descriptions serve to demonstrate that even though some aspects of an interface definition may vary, a basic requirement can still be met if the fundamental technicalities remain constant.

For most pre-launch satellite operations and tests, the associated electrical ground support equipments are almost adjacent to the satellite, often less than 500 metres away. This is a prime difference to the post-launch conditions, and needs serious consideration for pre-launch activity planning if some of the EGSE items are to be suitably representative of the post-launch situation. This matter is particularly relevant to the payload SCOE of a communication satellite as the location of waveguide test couplers, a form of umbilical connection, is most pertinent for RF interfaces. Indeed, the pre-launch activities associated with payload testing cannot be an exact simulation of post-launch operations. However, satellite operation and control tasks, if undertaken by the inter-segment control interfaces of telemetry and telecommand, do enable ground segment commonality to be subjected to a systematically engineered solution.

Before the naturally enforced remote operation of the spacecraft from the OCC, verification of OCC performance is made before launch. This has been described in Chapter 3 with a communication satellite being remote from the OCC. Although remote satellite operations from the OCC before launch can remain unchanged, developments of these fundamental techniques have been adopted and expanded to support satellite constructional tasks.

It can be advantageous to remotely operate a spacecraft for pre-launch testing purposes. The following description of such an application can be used as a demonstration to show the similarities that can be accomplished in pre- and post-launch remote operations.

It has been explained that the satellite and the OCOE are located in close proximity. However, activities associated with remote pre-launch operations have required the satellite to be located at a test site which is remote from the OCOE. To meet this requirement, development of equipment has been necessary so that at the satellite location safe and efficient telemetry and telecommand interfaces are established. This equipment, the OCOE remote workstation, has identical telemetry and telecommand equipments to those which are used in the OCOE. The remote workstation is equipped with local control and display facilities which allow the satellite to be adequately controlled should the connection to the OCOE be lost or impaired. The remote workstation connection to the OCOE is made via a dial-up duplex telephone line and modems. An overview of the most basic architecture for this remote OCOE application is shown in Fig. 6.4. The principle of this operation has been implemented by ESA on the ECS and OLYMPUS-1 programmes.

Communication is achieved with PCM-SPL signal liaisons being made at bit rates which can be safely handled by the duplex telephone lines. To cope with the constraints imposed by the duplex lines, intermediate bit rates were introduced for telecommands. Telemetry data were linked at the normal satellite bit rates, as the telephone lines did not present any constraints.

The use of telecommand bit rates for the AIT application within the ranges defined in Table 6.1 is achieved by adopting buffering technologies between the OCOE and the OCOE remote workstation. This has been made possible by modifying the rather simple ground station adaptor (GSA), which is used during OCC software validation testing, to become the OCOE Remote Adaptor (ORA) identified in Fig. 6.4. The interfaces between the OCOE remote workstation and the satellite do comply with the post-launch telemetry and telecommand bit rate requirements. The basic aspects for the remote control of satellites has been demonstrated to be identical, no matter whether the distance between the satellite and the telemetry — telecommand facilities is 36 000 km (ECS geostationary orbit) or 3.6 km (ECS pre-launch mechanical vibration tests).

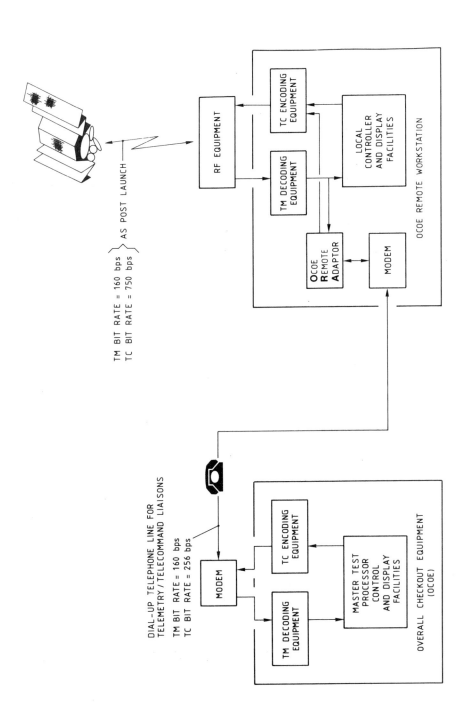

Fig. 6.4 — Pre-launch ground segment configuration for remote operations.

Table 6.1 — Telemetry/telecommand bit-rate utilization for remote OCOE operations

Satellite models	Intermediate bit rates for ORA application (bits per second)	
	TM	TC
ECS	160	200–300†
OLYMPUS-1	256	200–300†

†256 bps nominally

Such a demonstration should emphasize that common operational methodologies can be accomplished within the Earth segment applications of both pre- and post-launch activities. From an overall standpoint, therefore, provided that the system is designed with efficient modularity the changing of the telemetry and telecommand transmission media should not have a significant impact upon spacecraft operation.

6.3.2 A Post-launch re-use of pre-launch facilities

For the ESA communication satellite programmes, some facilities which were developed for pre-launch activities have been re-used for post-launch operations. These facilities are known as the communications satellite monitoring facilities (CSMF) and are employed primarily to assist with the fully operational on-station situations specifically applied to the satellites which have been constructed under the auspices of ESA.

The purpose of the CSMF is to give satellite design engineers and specialists easy access to engineering data transmitted from satellites during their construction, readiness for launch, launch, and orbital operations. The CSMF allows satellite status and short-term performances to be assessed in an on-line manner without disturbing the satellite in-orbit operations which are performed under the authorities who are responsible for the operations and control centre (OCC) activities. Similarly, the monitoring of any activity by the CSMF during satellite construction and preparation for launch is accomplished without disturbing the activities of the authorities responsible for satellite construction and test.

The overall architecture of the CSMF is shown in block form in Fig. 6.5. The configuration of the CSMF central facilities enable the OCOE user software which was utilized during satellite construction to be employed. The CSMF can be configured to display processed telemetry data in an engineering form by monitoring the housekeeping telemetry data received from orbiting satellites. Before launch, satellite telemetry in PCM form can be relayed over dial-up telephone lines from the satellite location to the CSMF location. The telephone lines connect the CSMF to the satellite EGSE where the satellite telemetry signal is available. The characteristics of telephone lines dominate the utilization methods for the relay of telemetry signals. These have not been restrictive for the housekeeping telemetry bit rates that have

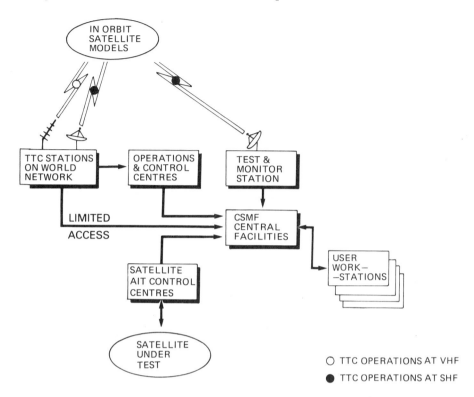

Fig. 6.5 — ESA communication satellite monitoring facilities (CSMF). Overall architecture.
(Courtesy of ESA).

been described. Furthermore, the development of communication networks will enable any restrictions which may occur to be eased in the future.

The CSMF data processing and distribution systems use the European test and operations language (ETOL). Reference to the *Appendix* may be helpful in appreciating the application of ETOL. The colour synoptic pictures are particularly relevant to CSMF activities. The ETOL configurations for pre- and post-launch monitoring activities are almost identical. A major difference is associated with limit definitions within the telemetry database. As explained in Chapter 5, if limits are exceeded then outputs are made to display devices. Temperature limits for pre- and post-launch operations can be expected to differ to some extent owing to environmental conditions experience by the on board equipment. For testing purposes during AIT, limits may be narrowly set to assess performance, whilst for operational purposes after launch, the limits can be relaxed.

The CSMF workstations allow ETOL colour synoptic pictures to be displayed, enabling satellite performance to be assessed in real-time at both local and remote locations. A remote CSMF user workstation can with existing technologies be thousands of kilometres from the CSMF centre. The colour synoptic pictures which are produced by ETOL illustrate spacecraft performance and are relayed to the CSMF remote user workstations where the user selects which pictures are to be displayed. Furthermore, users can have a control capability if a user workstation is equipped with a VDU and keyboard. In this case, the CSMF centre, whilst retaining control can delegate and give users increased flexibility, providing a controlled independence for operational activities.

The systems engineering technique of taking advantage of the commonalities within pre- and post-launch Earth segments of an aerospace system is not limited to space exploitation endeavours. The exploration of space can also make use of this systematic approach. Indeed, the scientific exploration associated with the GIOTTO spacecraft's encounter with Halley's Comet during 1986 is a good example of how space exploration can be achieved with established engineering technologies being combined with a probing mission into the unknown. The GIOTTO spacecraft made its closest encounter with the comet approximately eight months after its launch. Data were archived during this period so that a significant amount of off-line data processing could be accomplished. Such a process was necessary to allow the maximum knowledge to be gained. These operations were part of the spacecraft mission operations and control activities. The EGSE that was utilized during the construction of the GIOTTO spacecraft, primarily the SCOE associated with the scientific probing instruments, was re-used with the OCOE to allow the effects of the close encounter to be monitored as it happened and to give assistance to OCC activities. This post-launch configuration was given in Fig. 4.2.

It has been shown that components of ground segments can be made with a significant degree of commonality for pre- and post-launch operations. Furthermore, there could be technical rationalization aligned to spacecraft control methodologies for systems used during the pre- and post-launch phases.

6.4 COMPREHENSIVE RATIONALIZATION

Rationalization can be an ongoing process that standardizes operational working practices. This can entail forward technical planning which defines and employs new technical standards. In addition, a more efficient implementation of those which exist is also a possibility.

All the phases of a system programme from design conception, through construction, launch, and orbital operation deserve a systematic appraisal from start to finish. The design of a space segment, besides being undertaken for achieving the ultimate users technical requirements, should consider construction and operational aspects. These design aspects do merit consideration, even if they do not emanate directly from the users' requirements. That is, the space segment designs should encompass Earth segment requirements if an efficient effective aerospace system is to be produced.

6.4.1 Terminology considerations

Differences in terminology have underlined the need for a comprehensive rationalization process regarding pre- and post-launch satellite operations. For post-launch activities the terms Earth segment and ground segment exist alongside TTC station, ground station, and Earth station. For pre-launch operations, the terms checkout station, checkout equipment, and ground support equipment are prominent. For both pre- and post-launch activities, an emerging term for one of the common elements is workstation. A more precise definition of the elements of an aerospace system could help the rationalization process.

Adoption of the practical engineering definition, terminal, could perhaps be a better description for some user Earth stations, the communication payload test and monitor stations (TMS) referenced in previous chapters being a good example of this misnomer. Expanding developments of technology certainly allows for a user requirement that can be met by a mobile communication system. Such a user requirement will allow terminals to be mobile on land, sea, or in the air, reinforcing the misuse of the term station(ary) if it is adopted.

As satellite control is a significant part of both pre- and post-launch operations then elements of both Earth segments could be assigned the term spacecraft control system (SCS). For satellite operations before launch, the OCOE is the control of testing centre and the OCC is the control of utilization centre for post-launch activities.

The term workstation needs to be applied with caution. Mobility can be a requirement for a workstation, with working methods being interactive or of an observer status only. Therefore, station(ary) can be of relevance. Perhaps the term man–machine terminal would be more appropriate and cover all situations. The definition special checkout terminal may also be a more exact definition of an SCOE which does have workstation capabilities.

6.4.2 Satellite control activities

For a geostationary communication satellite, operational activities regarding control, testing, and utilization can be subjected to a stringent rationalization review.

From the analyses of pre- and post-launch control activities which have been described in previous chapters, the comparisons shown in Table 6.2 are relevant.

This comparison summary should enable common design criteria to be formulated and baselined for both pre- and post-launch operation of the space segment. Thus, a rationalization of items (1) and (2) in Table 6.2 should be possible for AIT and mission control activities, giving a common design for control centre elements.

Testing and monitoring systems are necessary after launch to ensure that the communication links can be commissioned and operated. When the satellite is positioned in its geostationary Earth orbit, testing activities are undertaken at regular periods. These in-orbit tests (IOT) are associated with the communication payload to check performance parameters. The IOTs commence with the commissioning task which is undertaken by a test and monitor station (TMS). A TMS requests the OCC to configure the satellite to the architecture required for the communication payload commissioning activities. After commissioning has been completed, users can then employ the facilities which are provided by the spacecraft. Referring to point (4) of Table 6.2, consideration can therefore be given to undertaking a single design exercise for test and monitor stations and payload SCOEs which operate under the executive of the OCC and the OCOE respectively. Indeed, perhaps a communication payload SCOE can become an element of a TMS after the communication satellite has been launched, especially if liaisons with the OCOE and the OCC are identical. The rationalization which deserves consideration for communication systems which employ a spacecraft is summarized in Table 6.3.

Table 6.2 — Pre- and post-launch comparison of activities

Satellite Activity	Pre-launch (AIT)	Post-launch (Mission Control)
(1) Health control	Full control and status checks carried out. (simulated operations)	Similar to AIT but possibly not so extensive, dependent on (2) below
(2) Redundancy/ recovery	These features are comprehensively tested	If the satellite operates normally in orbit, these features may not be utilized, otherwise similar to AIT
(3) Orbit control	Special performance tests to establish tracking criteria	Tracking and orbit determination/control
(4) Communications payload operations	Payload entities tested	Payload entities tested, commissioned, and operated

This baseline should be considered during the early phases of a comprehensive programme if managed correctly.

The application of optimum rationalization can result in common methodologies for operation of the satellite encompassing pre- and post-launch activities. Thus, identical Earth segment elements can give support to space segment construction and post-launch operations if designs include the total life cycle of a satellite.

The area of greatest commonalities can be associated with satellite health control. After acquisition of the geostationary operational orbit, the mission control task is centred around the control of the satellite's health and status.

6.4.3   Harmonizing control

Earth segment development should be aligned closely to space segment construction. This would enable the construction process of space and Earth segments to merge more closely. Such a merger could give a construction activity that enables a spacecraft launch to be less of a step function in a programme. This approach would be to integrate closely the space segment pre- and post-launch construction phases, and to align the use of Earth segment components for pre- and post-launch tasks. This may require duplications of equipment but they should be the result of a single design and development activity.

The technical development of the Earth segment, although evolving from the post-launch requirement, needs to be implemented in harmony with the schedule required by the spacecraft construction process. That is to say, from the pre-launch development schedule rather than the post-launch operation timetables. It is evident

Table 6.3— Rationalization for communication systems

Item	System	
	Pre-launch	Post-launch
1	OCOE	OCC
2	Payload SCOE	TMS
3	TTC SCOE	TTC stations

that the pre-launch Earth segment which supports spacecraft construction needs to be operational before the post-launch ground support facilities. In fact, the need date for the pre-launch electrical ground support equipment (EGSE) as a working system can be a few years before the launch of a spacecraft.

Often, the post-launch ground segment is not needed to be fully operational until the spacecraft launch date is imminent. There are some ground segment developments which are necessary to meet post-launch requirements that are best delayed until the launch date approaches. Some final flight dynamics activities which embrace orbit prediction and control can be a case in point.

The final constructional activities of a space segment occur at a launch or afterwards, when the operational flight path has been acquired. The transfer orbits, which are necessary for placing a satellite into a geostationary operational position above the Earth, can be regarded as being part of the space segment constructional process. These events establish the requirement dates for ground segment facilities, the TTC stations, and the operation control centre (OCC).

Before the fully operational status of the satellite is attained, there will be variations in the ground segment configurations to cope with the orbital paths that the satellite will follow. For all pre-launch activities, whether minutes just before lift-off or many months before, the pre-launch Earth segment architecture composition may change, but the concepts should be unaltered. In principle, the ground segment architectures before and after the satellite's launch can be identical. At least from a spacecraft control standpoint, pre- and post-launch ground segments do have commonalities of purpose. There will be differences of course, in pre- and post-launch Earth segments. For example, SCOEs are required for pre-launch activities, but not post-launch. Flight dynamics are necessary for orbital operations, but not directly for pre-launch activities. Nevertheless, an executive control centre and user interfaces are required for both pre- and post-launch activities. If spacecraft pre-launch testing is to be representative of post-launch operations then it should be possible for ground support equipment architectures for spacecraft health monitoring and control to be basically identical for pre- and post-launch operations. Fig. 6.6 gives an overview of a programme for satellite operation activities depicting how satellite control is a central element.

The rationalization of techniques and technologies for the ground segments should enable a single design and development exercise to be performed. Require-

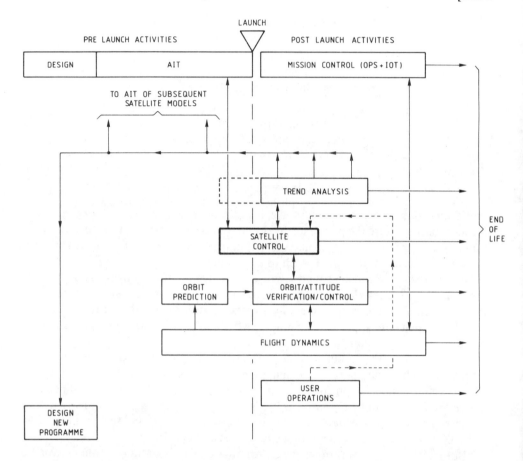

Fig 6.6 — Overview of a satellite operational activities. Satellite control is the centre piece for all operational tasks.

ments for both pre- and post-launch ground segments need to be considered. In fact benefits will be attained only if all efforts are effectively coordinated. Such efforts would give a reduction in errors as only one prime learning curve will need to be ascended, giving a minimum of repeated design and development efforts. From an overall aspect, schedules should be easier to maintain, thereby giving cost-effective results. This cost effectiveness could perhaps also be achieved by enabling post-launch operations to be more automated. Such a situation could be achieved by increasing the automatic applications for spacecraft health control. For launch and early orbit activities, this would not necessarily be an ideal solution, but after operations become a routine exercise, then such a move could be attractive.

The similarities between the system's requirements which are utilized to operate and test the spacecraft before and after its launch can be observed from what has been presented. There are differences in some operational tasks, but, in concept, the operations should be almost identical for the control of the health of the space segment. Thus, common designs could be made to suit pre- and post-launch purposes. The development schedules of both the satellite EGSE and the operational ground segments may demand duplications of hardware and computer software items. However, these duplications should be the result of one design and development process.

6.4.4 Rationalization of operational practices

The re-use of procedures that were developed and utilized during pre-launch operations could, with modification, be used for orbital operations. Such an approach of ensuring that the pre-launch operations of a spacecraft are a rehearsal for post-launch activities should result in working practices being harmonized. The formal procedures and associated computer software for pre-launch integrated system tests (IST) of the spacecraft could be re-used with modification to become the basis of flight operations procedures (FOP) that are employed for mission control activities. Certainly, the CSMF descriptions expounded earlier in this chapter have shown that computer software can be re-used. Therefore, the more that procedures are automated, the possibility for utilizing fully the commonality that exists within pre- and post-launch operational methodologies is enhanced.

It should be possible for some members of the spacecraft construction team to be transferred, in an executive capacity at least, to post-launch operations. This would ensure that spacecraft operation experts who gained experience of the spacecraft during its construction, actively utilize their experience at the post-launch operations and control centre. The senior engineer responsible for spacecraft construction could undertake the duties of an in-orbit spacecraft controller, or at least function as a most able assistant.

For geostationary communication satellites, the immediate post-launch activities are critical for the space segment's longer life. The construction team will be familiar with what is expected if the qualification programme, part of spacecraft AIT, has been correctly formulated. If this is the case, an unexpected situation occurring after launch could be handled efficiently as judgements and actions would be based upon hands-on experience. When the satellite is in a stable condition, the operation experts who were originally in the satellite construction team could be phased out of the orbital operations. They could then, of course, rejoin a construction team of a new programme whilst being available in a consultancy capacity for the orbital operations of a satellite they know well. The satellite operational tasks will, if all operates satisfactorily, be rather routine and mundane after commissioning activities have been completed.

In summary, it should be possible for a core team to be formed to handle spacecraft control and operation. The core team will need to be supported by design experts during spacecraft construction and by flight dynamics experts after launch. The size of the core team will be dependent upon the amount of expertise that is embodied in the system executive machines of the OCOE and the OCC. Thus, if computerized systems are adequately employed,the size of the core teams can be minimized as operations could be automated. The use of computer software which provides the same man–machine interface to the workstations of executive machines could enhance the management technique of efficient manpower utilization. Certainly, the tasks of establishing comprehensive procedures should harmonize and rationalize the operational practices for spacecraft during their construction and orbital life.

6.4.5   Matrix implementation

Comprehensive rationalization may be easier to accomplish if it is applied initially to a single and independent satellite system aligned to a particular programme. To some extent this activity is ongoing and can be considered as vertical rationalization. These principles are depicted for three independent programmes in Fig. 6.7. The establishment of vertical rationalization should help to enable common solutions to be implemented and re-used on an inter-programme and inter-discipline, exploitation/exploration basis. Such a process could be considered as horizontal rationalization. The concept of vertical and horizontal rationalization is shown in matrix form in Fig. 6.7 and should, ideally, be implemented accordingly.

With new avenues of development being opened, a consolidation of established techniques by a rationalization process will itself be a development field. System expansions associated with the exploitation of space should give due regard to utilizing fully the developments that have been made. Such an approach deserves consideration before embarking upon new developments which may not be strictly necessary if requirements are to be achieved with efficient solutions. However, this approach should not halt all development activities, some of which will be necessary for expanding both space exploitation and exploration endeavours.

To continue with what has always been the working practice by resisting change may not be the best method to achieve efficiency. Consideration needs to be given to the benefits which can be provided by changes that modern technologies supply. In summary, the consolidation and rationalization process should embrace stability but allow changes to occur.

The re-usability of pre-launch ground segment items within and between programmes should allow some items to be employed during post-launch activities. Indeed, the use of TTC standards paves the way for such an implementation which is being employed, to varying extents, upon different spacecraft programmes. Developments associated with rationalization should certainly enable dual development activities to be obviated in some areas, particularly in the spacecraft control field.

Internal space segment TTC subsystem interfaces have been subjected to some standardization within Europe. This internal space segment standardization has given the opportunity for telemetry and telecommand distribution to be placed upon an on board data handling (OBDH) bus. This has promoted a degree of spacecraft standardization between different programmes.

A standardization process can be applied to areas within and between pre- and post-launch ground segment TTC facilities, making the re-use of TTC SCOE a technical possibility. This exercise has resulted in the commonalities being highlighted and has optimized re-use. Telemetry and telecommand technologies can be used for pre- and post-launch activities as has been demonstrated practically for telemetry data by the communication satellite monitoring facility (CSMF) described earlier in this chapter.

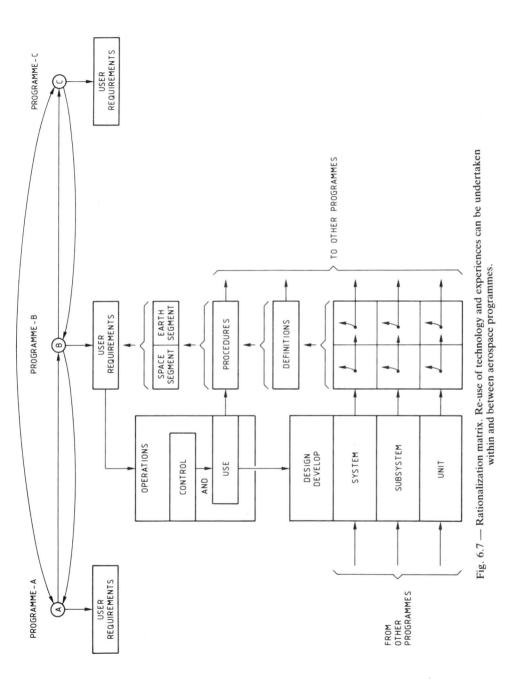

Fig. 6.7 — Rationalization matrix. Re-use of technology and experiences can be undertaken within and between aerospace programmes.

Other subsystem and system developments should be able to follow the process which has taken place with TTC subsystem construction and test. Spacecraft subsystem standardization could emanate from basic support fields, for example on board spacecraft power supplies. Similarly, the possibilities will exist for the standardization of other spacecraft subsystems and entities, together with their associated test equipments.

When spacecraft subsystem interfaces become more standardized, the opportunities will open for an increase in modular construction techniques to be employed. This will expand the possibilities for the application of system engineering techniques for spacecraft production and control. As spacecraft become more standardized, their construction and testing should enable the process to be operated on more of a construction line basis.

In conclusion, comprehensive rationalization should be a design requirement goal for Earth and space segments' architectures if cost-effective solutions are to be accomplished. A rationalization process aligned to spacecraft control employing existing TTC standards can be achieved to produce and operate more efficient aerospace systems. The expansion of systems will necessitate the use of existing standards and/or their enhancement. New standards will need to be developed so that the user requirements of evolving aerospace systems can be achieved. This evolution will embrace spacecraft control which should be conducted in a coordinated, coherent manner from a control centre.

7

Evolving systems:
retaining control

The evolution of aerospace systems will require developments to expand or replace those which exist as operational entities. Such developments should enable safe and efficient operations to be maintained if control centre methodologies are retained. Although the role of control centres can remain unchanged in principle, system evolution and expansion should give the possibility for increased delegation of authority. The implementation of this executive capability, if correctly applied, should ensure that spacecraft operations are coherently performed.

A balance of activities in differing development fields will need to be made as systems expand. The standardization of technology should be balanced with new developments which are undertaken to improve technical performances. Indeed, the most advanced technology should provide an elegant simplicity, at least to system operators and the ultimate users.

The application of the techniques which have been described can enable systems to evolve in a controlled manner if ultimate users or their representatives take part. Their participation may be limited to a consultative capacity.

Improvements in technical performance should embrace the system operators' requirements. The remote operations which have been described in previous chapters have been a step in this direction. Application of the techniques which have been developed for operations and control centre (OCC) compatibility tests and overall checkout equipment (OCOE) remote workstation utilization have employed a communication satellite when making liaisons between Europe and South America. The use of any communication satellite in this instance was transparent to the OCOE and OCC operators because the commercial networks which were employed did not require any special arrangements to be made. Nevertheless, even if more complex networks are employed and arrangements need to be made with commercial authorities, this should not affect operations at the workstation interfaces. Indeed, the utilization of a dial-up telephone line for these satellite pre-launch operations does emphasize this point.

7.1 SYSTEM EXPANSIONS

The space segment of an aerospace system will expand with continued developments and result in an increase of segment elements. These space segment elements will be manned or unmanned spacecraft and space stations. Liaisons will be undertaken between the space segment elements in addition to those with their associated Earth segments. Such liaisons will most likely be made via a communication satellite which will relay data between its users and provide a link for the control of space segment elements. This communication network can be known as a data relay system (DRS). Such a communication system has been employed for flights of the NASA Space Shuttle. Developments are ongoing within Europe so that evolving aerospace system requirements can be achieved with the use of a data relay system.

Existing and new locations of Earth segment elements, and differing spacecraft flight paths, bring about the need for an increase in communication links to enable operations to be undertaken safely. Some spacecraft engaged upon missions in near-Earth orbits can require a constant liaison with their ground segment. Spacecraft in elliptical orbits can be out of sight of their control centres for some periods. Thus, to maintain constant liaisons would require the utilization of several TTC stations. An alternative solution which is being developed uses a data relay system (DRS). Three satellites in geostationary orbits could allow constant liaisons to be made between space and Earth segment elements of an aerospace system having satellites whose orbits give interruptions in inter-segment communication links. The architecture of such a system was foreseen and documented in 1945 by Arthur C. Clarke.

An overview of a data relay system communication network that employs single communication satellite is shown in Fig. 7.1. In the relatively short term, communication between Earth and space segments of independent aerospace systems will be necessary. In the longer term, what is at present considered to be an independent aerospace system will become an element of an entire system.

7.1.1 Spacecraft cluster philosophy

Space segments composed of multi-elements which are unmanned have been produced. Such aerospace systems were constructed in the mid 1980s for space exploration and exploitation endeavours. Six spacecraft were engaged upon probing missions into the unknown environment surrounding Halley's Comet. There were no direct spacecraft-to-spacecraft liaisons, and each spacecraft was part of an individual and independent aerospace system engaged upon different tasks. However, the user scientists cooperated closely throughout the scientific task. From a technical standpoint, the space segment of the scientific probes was not closely bound but nevertheless, external aerospace system interfaces at the user end certainly were, as worldwide inter-programme support was achieved. The results of spacecraft encounters were exchanged and information from the other probes aided the GIOTTO spacecraft flight path activities. An overview of the scientific probes into Halley's Comet is shown in Fig. 7.2. These scientific probes around the comet practised in a fundamental form some of the principles that are applicable to the operation of a spacecraft cluster. The term spacecraft cluster applies to a group of spacecraft in close proximity engaged upon a common and coordinated task.

The fundamentals of spacecraft cluster control and operation have also been practised by space exploitation applications. The European communication satellite

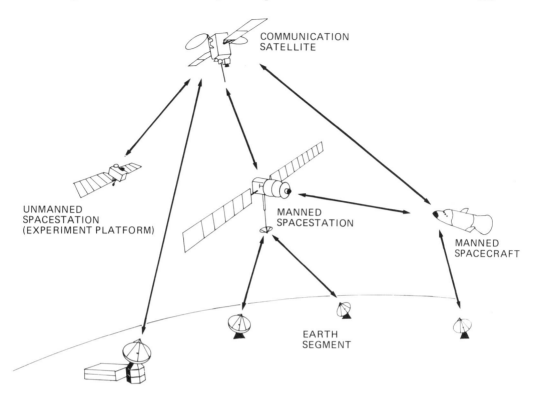

COMMUNICATION
SATELLITE

UNMANNED
SPACESTATION
(EXPERIMENT PLATFORM)

MANNED
SPACESTATION

MANNED
SPACECRAFT

EARTH
SEGMENT

Fig. 7.1 — Communication network for space segment expansions. The communication
satellite of the data relay system (DRS) will be a node which allows communication to be made
between elements of space and Earth segments (courtesy of ESA).

programme enables the space segment of the EUTELSAT communication system to
be composed of four ECS models which are in geostationary positions around 10
degrees east. These satellites do not have inter-satellite links (ISL), but the operation
of the ECS models, which have been renamed EUTELSAT I−1, −2, −4, and −5
after their successful launch and commissioning, is made from a single ground
segment operations and control centre located at Redu, Belgium. It is at this location
where the control operations of these satellites are linked and coordinated. Future
clusters of communication satellites with inter-satellite links are a possibility. Such
links could be utilized for redundancy and system operational purposes, but,
nevertheless, a single control centre should be retained to enable all operations to be
efficiently performed and coordinated.

Communication clusters of the future, shown in Fig. 7.3, could consist of a
number of satellites closely positioned around a nominal location in geostationary
orbit. All the satellites of the cluster lie within the beam width of all Earth station
antennas. The satellites are not physically connected, but are interconnected for
control purposes, at least via RF links, through a central control satellite.

Common to all configurations shown, the telemetry, tracking, and command

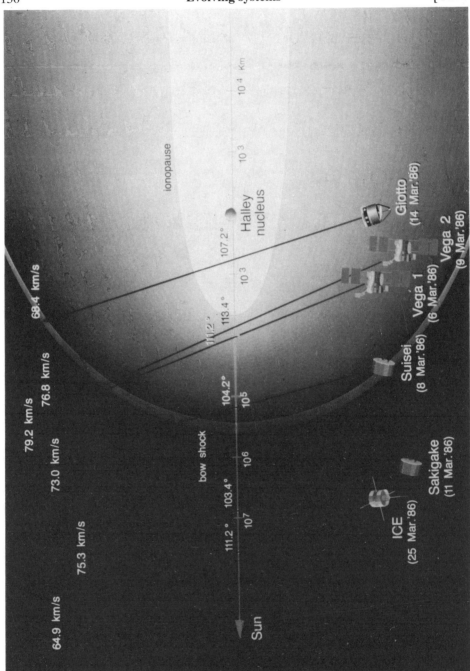

Fig. 7.2 — Spacecraft cluster: probes into Halley's Comet.
GIOTTO — European Space Agency (ESA); VEGA 1 AND 2 — Intercosmos of USSR
Academy of Sciences; SUISEI — Japanese Institute of Space and Aeronautical Science
(ISAS); SAKIGAKE — Japanese Institute of Space and Aeronautical Science (ISAS); ICE —
American National Aeronautics and Space Administration (NASA). The six spacecraft flying
through the coma of Halley's Comet at different times and at different distances from the
nucleus (courtesy of ESA).

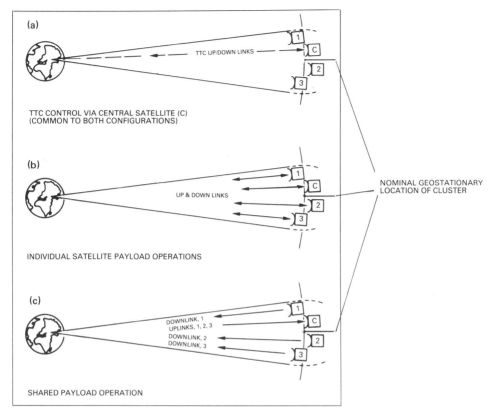

Fig. 7.3 — Operation of communication satellite clusters. A central control satellite performs
an executive function for two cluster configurations (courtesy of ESA).

functions are performed via inter-satellite links (ISL) through the central control
satellite (C).

Fig. 7.3 (b) shows the concept of payload module capacity sharing within the
cluster. Each satellite receives and transmits a channelized share of the total system
capacity. The configuration of each individual satellite does not differ markedly from
a conventional configuration except for inter-satellite liaison requirements. Communication uplinks and downlinks are achieved through each individual satellite, a user
telecommunication channel 'traversing' only one satellite.

Fig. 7.3 (c) shows the concept of communications service shared across a number
of satellites. The central control satellite (C) receives all the communication signals
from the ground, performs on board signal switching and routeing, then transmits
the signals through one or more subordinate payload satellites to the Earth stations.
This concept facilitates a gradual increase in communication capacity and the
replacement of failed satellites by new ones. This is achieved by adding or replacing
individual satellites commensurate with the interconnection capability of the central
satellite, and by replacing a low capability central satellite by a higher capability
satellite when required.

Basically, each of the satellites in such a cluster are conventional in type and, hence, has all the usual properties. However, the satellites will carry on board units to cope with rendezvous control, inter-satellite links, and accurate inter-satellite distance keeping.

7.2 STANDARDIZATION FOR OPERATIONS

Over the years, since the exploration and exploitation of space started, both the space and Earth segments have been constructed to meet specific requirements. This has resulted in a degree of formal technical standardization being achieved. These standards are particularly aligned to inter-segment interfaces for spacecraft control telemetry, tracking, and command (TTC). These interfaces between space and ground segments have enabled a set of uniform techniques to be established, allowing aerospace systems to be operated reliably, efficiently and with safety. European and American standards have been produced in a coordinated manner, enabling a significant amount of compatibility to be achieved. Furthermore, wider international cooperation is now established for further development of standards for the aerospace industry.

The most dominant standards for the operation of an unmanned spacecraft are the space/ground/space inter-segment interfaces which provide the spacecraft control capabilities. Advances in technologies will enable more efficient aerospace systems to be developed and controlled during the evolutionary process. Such activities will result in changes to inter-segment interfaces. These interfacing standards for telemetry, tracking, and command (TTC) liaisons will be a development from those which were formally established in the late 1970s. The general requirements which will be placed upon pre- and post-launch ground segments will remain identical in principle. That is to say, safe and efficient aerospace system operations will be implemented with tracking, telecommand and telemetry interfaces to support the spacecraft operation and control task. In addition, requirements for the Earth segment elements to utilize the space segment to meet the ultimate users' requirement is an area where continued rationalization and consolidation can establish the need for further standardization. The developments of aerospace system user requirements are undergoing change. The number of users is increasing, as are their specific requirements. The ultimate users of some exploitive endeavours and the ultimate users of scientific probes on spacecraft can benefit from more autonomous and flexible operations. However, this will be dependent upon operational methodologies.

The on-going developments of aerospace systems will necessitate the use of existing standards and their expansion so that a space segment can be operated and controlled to meet users' requirements. The application of the PCM telemetry and telecommand standards that have been described can be restrictive for some users. Spacecraft designs utilizing the time division multiplexing (TDM) techniques specified in the ESA PCM telemetry standard do have a significant influence which can be rather constraining. Changes to channel sampling rates cannot usually be made after spacecraft developments have terminated and operations commence. Data channels are sampled sequentially in a fixed cyclic manner with TDM according to the PCM telemetry standard described in Chapter 4. This method of source data collection can be satisfactory for the spacecraft housekeeping telemetry but it can present restrictions for the scientists engaged upon the analysis of experimental information. Experimental information from the on board spacecraft experiments normally necessitates fast sampling of telemetry channels. This can require telemetry bit rates in the order of 50 000 bps, whilst housekeeping bit rates of 400 bps can be adequate; this was the case for the GIOTTO spacecraft. Such a requirement can impose restrictions for users when a spacecraft employs the established PCM telemetry standard.

Summarizing, the application of established technologies which emanate from the PCM telemetry standard result in information associated with specific subsystems being scattered within a complete telemetry format and sampled at a defined and fixed rate. Whilst this TDM method for on board data collection has been employed satisfactorily, there are disadvantages, as explained. For more complex satellites a less rigid technology is required. This subject is discussed in the following sections.

7.3 PACKET STANDARDS

Developments associated with spacecraft operation and control, together with parallel developments associated with communication technologies, are resulting in the emergence of packet standards for telemetry and telecommand. The principles which are provided and have been established by the PCM telemetry and telecommand standards remain. The utilization of the developing inter-segment standards should enable the ultimate users of aerospace systems to be delegated, increasing independence from an executive controller (man or machine). This situation can be aided with the adoption of layering techniques by the implementation of packet telemetry and telecommand standards.

7.3.1 Layered message structures

Modern technologies concerned with data transmission systems are developing with the establishment of layered message structures. This method allows the collection and transportation of data to ultimate users to be achieved in a more flexible manner with safety. The rules governing the data exchange between the layers can be fixed to a standard defined method of operation. This means that each layer has an exact interface with the layers above and below it. Therefore, as long as the rules governing the liaison with adjacent layers are obeyed, there should be no significant constraints on the internal operations within a layer. This approach is a modular structured data system which can be developed and adapted to meet evolving user requirements.

7.3.2 Packetizing principles

Packet telemetry facilitates the selection and routeing of data more directly by and to the ultimate user during spacecraft operations. Packet telecommands, however, although they may originate directly from various parts of the Earth segment, should be routed with executive control for authentication purposes if control centre concepts are to be continued.

To enable a comparison of how the packet standards can be considered as an expansion of established PCM telemetry standards, a further brief examination of operations utilizing the established PCM telemetry standard will be useful. Consider the task of collecting analogue source data on board a spacecraft and transmitting them to a user within the Earth segment. This task, in practice, consists of a number of sub-tasks which are undertaken by units of hardware and software modules. The subsystem tasks associated with telemetry data on board a spacecraft are now outlined in a sequential manner:

(a) Collect the analogue source data from on board sensors.
(b) Convert analogue values to digital TM values.
(c) Insert the TM values at the appropriate place in frame and format (TDM).
(d) Add information for identification purposes (frame ID, format ID, spacecraft ID).
(e) Add synchronization code.
(f) Encode the data into a PCM signal.
(g) Pass on the data, which are now in PCM signal form, for modulation and transmission.

It can be seen that this method of data transmission to some extent, packetizes information for transmission. Although it was never formally described as such, tasks (c), (d), and (e) provide a form of communication protocol. In fact, the principles of constructing information into telemetry packets is practised to some extent by the established PCM methodology of spacecraft TTC subsystems. This procedure then has to be reversed by the Earth segment elements which receive and monitor telemetry data.

A similar sequence of events is undertaken when operations are performed in accordance with packet telemetry standards. To facilitate a better understanding of data packetization, these basic operations which employ the ESA packet telemetry standard will now be considered in more detail.

The fundamental criteria for the collection of information (telemetry channels) from a spacecraft subsystem or experiment and the delivery of these data to the ultimate user may be summarized as follows:

* The frequency of data sampling should comply with users' needs which can vary, owing to operational requirements.
* The source data packets should be readily identifiable to enable easy extraction and transportation to users.
* The system should be such that errors caused by the transmission media can be identified and corrective actions be undertaken.

The general sequence of events for telemetry data flow between the data sources on board the space segment and the ultimate users located within the Earth segment are shown in Fig. 7.4. A brief description of the data flow follows, and more technical details are defined in the relevant standard (refer to the *Bibliography*).

The figure shows that telemetry packet operation employs two prime data communications layers:

* User data layers.
* Transport layers.

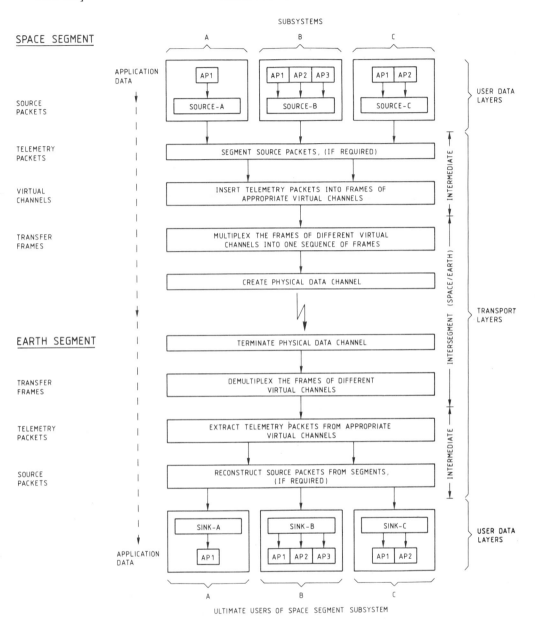

Fig. 7.4 — Packet telemetry data flow. The process of transporting user data from the space to the Earth segment.

The space segment user data layer is operated under the executive of the ultimate users of on board subsystems. The ultimate user can normally specify the form of telemetry packets which are relevant to the associated subsystem. These subsystem operations can be conducted in real-time. This flexibility will have been built into the subsystem during design and development. Source packets will contain information (telemetry channels) from an experiment or subsystem. This information is termed application data. In general, the user should be able to optimize the size and structure of application data sets. The user will therefore be able to define a data organization relatively independently from other users, and adapt its organization to the various modes of operation of the associated subsystem data sources. The organization of data sources need not necessarily be constrained to single subsystems as indicated in Fig. 7.4. Application data could be obtained from several subsystems to form telemetry packets. The user control of telemetry packets can be direct via telecommand or indirect via the status/condition of the relevant spacecraft subsystem. This latter control of telemetry packets would be autonomously performed by the on board subsystem.

The transport layers facilitate the application data transfer from space to Earth according to defined communication protocols. A telemetry packet provides the intermediate transport protocol. A transfer frame is the data link protocol data unit. The transfer frame is used to transfer the source data across the inter-segment interface of the space and Earth segments. The transfer frame also provides the methodology to time-share the data link between groups of sources by creating logical virtual channels. The term virtual channel is used to differentiate between this and a real channel which normally has a hard physical connection. The protocol allows the telemetry packet to contain segmented source packets. This avoids the seizure of the virtual channel by very long source packets. The transfer frame structure enables frames of different virtual channels to be multiplexed into one sequence of frames for transmission by the physical channel.

Fig. 7.4 also shows the physical layers associated with the inter-segment interface. These layers connect:

- the transfer frame with transmission equipment, and
- the reception equipment with system components which eventually present application data to users.

In summary, the implementation of packet telemetry requires a significant amount of data to be transmitted, of which a large amount is the protocol that provides flexibility and safety to the data of ultimate users. This can be considered as an overhead for the conveyance of applications data which can be offset by the increased on-line control of a spacecraft subsystem by an ultimate user.

7.3.3 Packet applications

Significant autonomous operation of space segment components by ultimate users will be aided by the application of packet standards and expanded performances of communication systems. This autonomous operation is shown in a hypothetical form in Fig. 7.5. Information from spacecraft payloads (P/L) is relayed to users via a communication satellite which is controlled from its own OCC (not depicted on the figure). The overall control of the spacecraft is maintained from its OCC. Packet standards can enable telemetry data to be distributed in a more autonomous manner, thus enabling ultimate users to have more direct control of their in-flight payloads. Nevertheless, spacecraft control should be maintained under the executive of the

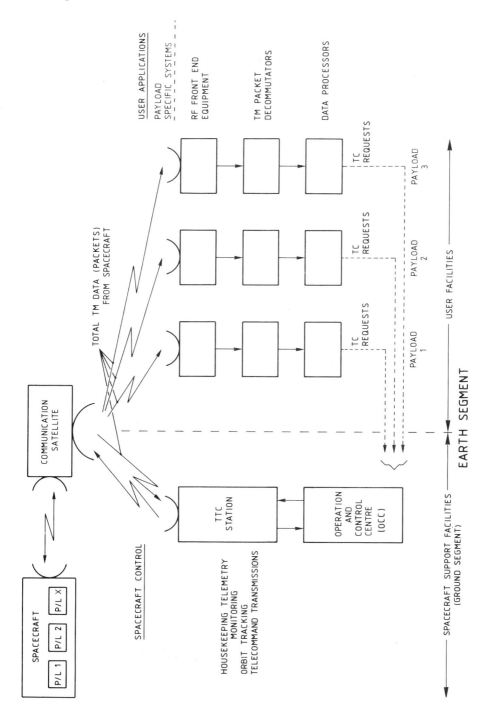

Fig. 7.5 — Packet telemetry application. This is a hypothetical example of a closed-loop configuration which may be implemented as aerospace systems evolve. User facilities may be delegated increased autonomous control by the OCC.

OCC. Such a situation will be possible as the necessary system elements are developed. The RF front-end equipments will become smaller and, eventually, financially attractive as numbers increase because of expanding requirements. Therefore, provided that the users' antennae are within the communication satellite antenna footprints, reception of data as shown in Fig. 7.5 should be possible in the not too distant future. A telemetry packet decommutator will be necessary to provide the ultimate user with the specific data that are relevant to the associated in-flight payload. In the example shown in Fig. 7.5, the user front-end equipment will receive all telemetry data. A user packet decommutator will extract the relevant data and make them available for processing. The processing of the payload data will be undertaken by individual data processors which can, possibly, be undertaken by personal computers.

Telecommand requests will be routed to the OCC until technology and packet telecommand operations are sufficiently developed to allow direct transmission to the spacecraft. While the current situation prevails, telecommand requests from users to the OCC can be achieved via commercial networks. The control centre will then filter these user requests to avoid incompatible telecommand actions being executed on board the spacecraft. When the techniques associated with packet telecommand standards are adequately developed, filtering of telecommands will be possible on board the spacecraft, and not at the OCC as depicted in Fig. 7.5.

The advantages given to an ultimate user when the technologies associated with packet telemetry are employed can be summarized as follows:

- Telemetry data relevant to user applications have an individual address code which should enable telemetry packets to be readily identified and extracted.
- The sampling rates of telemetry data can be modified to suit user requirements.
- Errors caused by the transmission media can be identified and corrected.

The similarities of systems that operate with PCM or packet telemetry standards can be seen by comparing Fig. 7.5 with Figs 3.1 and 3.2 of Chapter 3. The extreme interfaces of the overall system are not modified in function by the changes from the PCM telemetry and telecommand standards to the packetized approach. With packetization, system level overheads can be incurred and should be evaluated before adopting these standards. The foregoing explanations give an insight into the benefits for system level operations which need to be examined and balanced against overheads before implementation is undertaken.

For spacecraft control purposes, the requirement will remain for encoding of on board information (analogue and digital parameters) and its eventual presentation to the executive controllers and users. Packet standards can aid the liaison and presentation of this information from one extreme to the other.

7.4 EVOLVING OPERATIONAL METHODOLOGIES

With the advent of increased efforts in space and with the development of increased international cooperation, standardization can be expected to continue evolving. This will impact upon spacecraft operational methodologies. Certainly, this will need to be the case for manned flight spacecraft and space stations, with authority being delegated from the ground to space segment elements.

Operational methodologies for unmanned Earth orbiting satellites will be influenced by the technologies that exist. These can exhibit drawbacks for applications to expanding aerospace systems. Developments will most likely evolve to include the following aspects.

7.4.1 Closed-loop aspects

Possibilities exist for making closed-loop responses for spacecraft control within ground segment facilities. The analysis of telemetry data and subsequent transmission of telecommands can be automated. To enable closed-loop responses to be made in shorter times, a move away from techniques that multiplex all subsystem telemetry data within a single telemetry format is a possibility that is presented by packet techniques. In addition, the possibility of generating and transmitting telecommands from several independent locations where the associated telemetry is processed is becoming feasible. Such opportunities will be possible with the adoption of packet standards for telemetry and telecommand.

Another route is to put some of the executive controller responsibilities for spacecraft health on board the spacecraft, thus reducing or obviating the need for telemetry and telecommand liaisons between space and ground segments for this control purpose. No matter what course is followed, executive software will be required, but where this resides will depend upon the system requirements. The assigning of intelligence to on board satellite subsystems and increasing automation for satellite health control can reduce the work load on the ground segment which may, however, retain the master role whilst delegating activities to the space segment. Certainly, until the role is delegated in an autonomous manner, some monitoring and control should be maintained by the master executive of the ground segment. Indeed, in the first instance, ground segment OCC activities could be more automated before such control is delegated to on board satellite subsystems. After this development has been successfully implemented, these tasks can be embraced within onboard spacecraft designs if this is shown to be advantageous.

Closed-loop operations of the aerospace systems described have executive control residing within the ground segment. The computer software for the executive machine can enable operations to be automated. The automation could make the man–machine interface simpler if it is efficiently implemented. Such endeavours may necessitate an upgrade to established executive control facilities for some spacecraft programmes. This could result in the production of differing versions and derivatives of ETOL to satisfy particular purposes which the existing version of the language cannot perform. Such developments should embrace a review of technologies that are directly associated with spacecraft operation and control.

7.4.2 Test aspects

From a more general standpoint, test facilities for pre- and post-launch activities deserve attention. Indeed, the requirements of pre-launch assembly integration and testing (AIT) should be comprehensively evaluated during the early phases of a programme. For example, the necessity of having umbilical connections on a spacecraft for pre-launch AIT purposes should be reviewed during the design phase of the complete system. Umbilical test connectors and associated cabling contribute to the weight of a space segment and, consequently, to the mass that the launch vehicle will have to lift. Testing methods can reduce or increase the requirement for umbilical connections. Therefore, the use of umbilical connections for spacecraft pre-launch operations requires consideration in the early phases of a programme. An alternative to umbilical connections could in some areas be achieved with telemetry channels which are used only for AIT purposes; this could be a packet telemetry application.

If the post-launch ground segment is developed in an insular manner, complex space segment simulators can be required. Comprehensive system designs do need to consider the true requirements for pre-launch simulators. Simulators may be necessary to test failures and automatic recovery modes of post-launch control software for the OCC. This cannot be tested with a 100% working spacecraft in AIT. Therefore, for some instances, simulators are certainly necessary. A coordinated systematic approach could ensure that the need for simulators is reduced. It may be possible for cases to arise which make the use of simulators unnecessary for some purposes. As was explained in Chapter 3, the operations and control centres have been commissioned as a pre-launch activity for most European communication satellites, utilizing the satellite they control after launch. This can reduce the need for satellite simulators to some extent, although if independent design and development activities are undertaken for pre- and post-launch spacecraft operations, there will remain a definite requirement for complex simulators for software development purposes for the post-launch OCC.

7.4.3 Test and launch aspects

Construction activities can require the transportation of spacecraft to various sites. This can be time consuming and include pre- launch Earth segment movements and commissioning. Some of these activities, as stated in Chapter 6, can be undertaken without the movement of some EGSE items. However, in some cases, when a spacecraft needs to be transported, this can be used to advantage. The ride a spacecraft experiences upon its launch vehicle can be more demanding than its journey from its construction site to the launch site. Therefore, if the spacecraft is adequately tested before its departure to the launch site a health check on arrival should be sufficient to enable pre-launch preparations to commence. It will then be necessary for only a small team of personnel and a minimal amount of ground support equipment to cope with such launch preparations. If this is achieved then the overall checkout equipment (OCOE) could be employed to become the post-launch ground segment executive controller of the operations and control centre.

It can be expected that satellite launch sites will be expanded. These expansions will not necessarily be limited to extensions of existing sites, but include the construction of new facilities at other locations. Indeed, launch site locations can aid the flight paths of launch vehicles and spacecraft. The use of different geographical sites can therefore have an influence upon the launch vehicle design and operation.

The development of launch vehicles are now undertaken from a more commercial standpoint rather than the experimental developments that brought about the current status. Development activities are based upon established practices, which is a systematic approach to continued improvements in performance. Such developments are presented in Fig. 7.6. This shows how progressive developments from Ariane 1 to Ariane 4 have enabled payload masses to be increased.

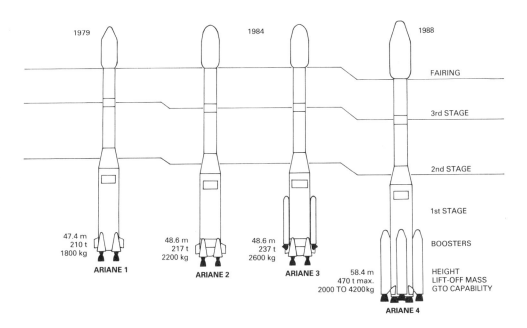

Fig. 7.6 — Developments of the European launch vehicle Ariane (courtesy of ESA).

7.4.4 Workstation aspects

Workstations are necessary for pre- and post-launch spacecraft operations. Advances in closed-loop operations should bring OCC workstations into a more passive mode for on-station geostationary communication satellites. This status will be dependent upon an increase in the automation of satellite operational methodologies.

Pre-launch testing activities for spacecraft will require an interactive man–machine dialogue to be maintained for checkout activities. SCOE developments can enable more autonomous activities to be carried out by these systems during AIT. For efficient and coherent testing to be performed during spacecraft system level AIT, control centre facilities should be maintained with an executive terminal. Indeed, workstations should be terminals which provide the user-friendly man–machine interface to suit the purpose of satellite control. To enable this to be satisfactorily achieved for evolving operational practices can, as previously stated, require computer software developments to be undertaken.

Any such actions should consider the following when advocating change:

- Maintenance of compatibility with proved and established techniques.
- The user interface to be kept to a minimum complexity.

The evolution of operational practices for satellite control should embrace workstation requirements that includes both pre- and post-launch aspects.

7.4.5 Safety aspects

System evolution will result in the advent of increased space flight and space travel to space stations. Certainly manned environments will require system developments to be engineered with diligence if safety is to be regarded as a necessity. Such systems certainly should be constructed with the knowledge gained. This can result in the consolidation of developments in various fields so that spacecraft of future aerospace systems can be controlled in a comprehensive manner during construction and utilization.

Manned flights of the future will employ data relay systems that embrace communication satellites to an increasing extent. Therefore, safety will certainly embrace a human aspect making safety a direct factor of efficiency. To a certain extent the safety factor has been taken into account with most European communication satellites that have provided a service to the general public. Autonomous actions on board some satellites have enabled unexpected situations to be handled by placing the satellite into a safe mode of operation which can interrupt services. However, recoveries have been possible under manual operations from ground segments to enable a resumption of services.

A more direct re-use of pre-launch AIT technologies and techniques for post-launch operations should make post-launch ground segment validations unnecessary. Therefore, post-launch activities should be subjected to less error and, consequently, should be safer. Furthermore, for a geostationary communication satellite, the on-station activities could benefit from the automatic operations that are practiced during AIT. This would relieve the pressure on satellite controllers, reduce the manual activities that are employed on a monitoring basis, and contribute to safe operations.

7.5 ADVANCED AEROSPACE SYSTEMS

Looking to the longer term future and considering space stations as manned living and working environments, it will certainly be necessary to examine how such systems will operate from social aspects in addition to technical standpoints. In such an environment, the working situation and social scene will be very close. Certainly, some basic social and technical standards will be necessary for international space ventures, with language and technical acronyms being of particular relevance. This can be practiced on the Earth to some extent before the space segment station is fully operational. The environment of modern submarines has a similarity with manned spacecraft and space stations. The exterior environments of a submarine and a spacecraft are hostile to human life, thus systematic endeavours can be aligned within terrestrial and non-terrestrial experiences.

7.5.1 An example

The operational scenario of an advanced aerospace system which employs a satellite in near Earth orbit is shown in Fig 7.7. This spacecraft is an unmanned retrievable platform that can be used for space exploitation and exploration endeavours. The ESA EURECA (European retrievable carrier) programme will produce the spacecraft and as will be seen from Fig. 7.7, launch and retrieval will be undertaken with a space shuttle. During the full operational phase with experiments being operated,

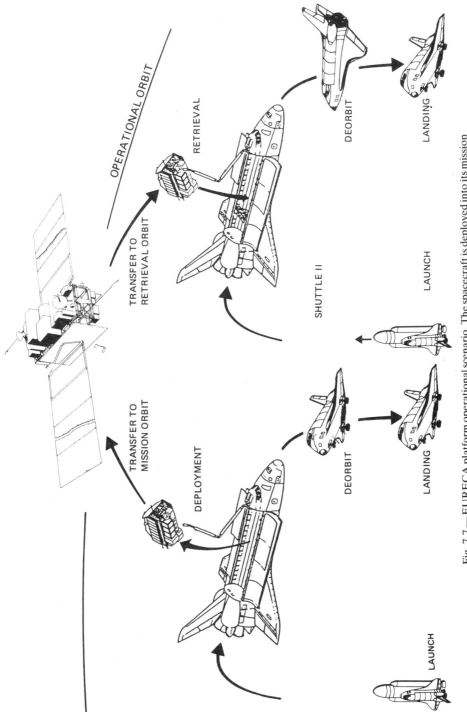

Fig. 7.7 — EURECA platform operational scenario. The spacecraft is deployed into its mission orbit using a space shuttle. Some months later, a space shuttle will retrieve and return the spacecraft to Earth (courtesy of ESA).

EURECA will be in an elliptical Earth orbit. This will constrain the real-time contact with the EURECA dedicated ground segment. The use of a data relay system that uses a geostationary satellite would extend the real-time contact with the spacecraft.

EURECA will be able to maintain contact with its Earth segment for extended periods by employing a data relay system. Initially this will be accomplished with a single communication satellite that utilizes K-band links (KA 20–30 GHz). An inter orbit communication (IOC) subsystem will be installed on the EURECA platform. This will allow communications to be made with the Earth segment via the 20–30 GHz payload of the OLYMPUS-1 satellite. An overview of this system is given in Fig. 7.8. The test station/terminal that was employed for the in-orbit tests

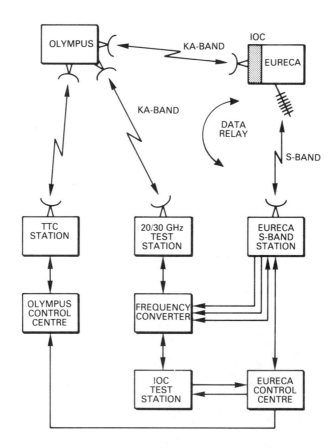

Fig. 7.8 — Data relay system for EURECA operation. A comprehensive use of ground segment equipment (courtesy of ESA).

(IOT) and demonstrations associated with the 20–30 GHz OLYMPUS-1 payload will be re-used as the front-end of the facilities necessary for commissioning the IOC subsystem. Frequency converters will allow the IOC test station to be a re-use of special checkout equipment (SCOE) that was employed during EURECA pre-launch activities. As EURECA utilizes packets techniques for telemetry and command purposes, then after the IOC subsystem has been commissioned, user activities could be undertaken along the lines outlined earlier in this chapter.

The EURECA platform which is supported by a ground segment during construction and launch, will require support during post-launch operations. After retrieval and landing, the platform will undergo tests and refurbishments before another flight. Therefore, the controlling function of the ground segment will be increased and make re-use an even more attractive proposition for pre- and post-launch operation of this advanced spacecraft.

7.5.2 Operational considerations

Advanced aerospace systems will employ communication satellites which are operated in Earth orbits. These could be of the geostationary type and also could be non-equatorial and elliptical orbits. As described, space segments of an aerospace system that meets a communication requirement will not be fully operational until in-orbit tests have been successfully completed. Therefore, satellite pre-launch testing activities deserve much more consideration if safe and efficient systems are to be produced.

If pre-launch activities are constrained to a build and test approach that does not embrace full post-launch simulations to a significant extent, then, in such a case, operations and ground segment performance are not practised as a rehearsal of post-launch activities. Consequently, a re-use of operational procedures and supporting computer software is not such a candidate for vertical rationalization within a programme that produces an operational satellite. However, with horizontal rationalization the re-use of technologies between programmes for pre- and post-launch aspects remain.

It is deemed that satellite operations commence before launch. This is due to satellite system level AIT activities embracing simulations of the launch and the space environment. These simulations have included environments which have been representative of the:

- expected conditions which occur during launch vehicle operations
  - ★ vibration tests;
- satellite post-lunch manoeuvres and on-station operations.
  - ★ thermal vacuum tests

In such a case the integrity of the satellite can be exercised in all modes of operation, including recovery actions from the unexpected and the exercising of redundancy. These activities could be considered to be an evaluation of a satellite's own safety and the service it provides to its users.

Pre-launch technologies can be employed for post-launch activities. Their implementation could be expanded to give a more direct re-use within a satellite programme, enhancing vertical rationalization besides horizontal movements between programmes. If such actions are undertaken, then a computer software kernel can be employed. It can be developed and operated before launch, then modified, edited, and supplemented for post-launch activities. Modular software designs, including the kernel which will be predominantly associated with satellite control, can be independent of hardware. This, however, will be strongly dependent upon a layered approach to software design coupled with a modular implementation of hardware requirements. Therefore, the software can be transportable and, not being tied to hardware, ensures a long life for its applications. As stated previously, software updates will become necessary, culminating in the necessity for rewrites. However, a re-use of hardware and software for satellite operations can be possible over a period longer than a decade.

Such an approach of applying the re-use of techniques for the evolving satellite operations methodology should ensure that flight paths are safely and usefully employed, with users' requirements being met in full. In summary, a systems approach to the design and control of aerospace system operations in both scientific and application categories will enable requirements to be efficiently achieved.

Bibliography

The documents of this Bibliography are not directly referenced in the text of the chapters. They are listed in the general order which they apply, and there is a significant amount of cross-applicability throughout the book.

(1) ESA BR-08 — ECS data book
(2) ESA PSS-46 — PCM telemetry standard
(3) ESA PSS-45 — PCM telecommand standard
(4) ESA PSS-04-106 — ESA packet telemetry standard
(5) ESA EGSE documentation — refer to document chart

The ESA EGSE documentation chart shows the documents which are produced under the auspices of the ESA European Space Research and Technology Centre (ESTEC) to aid the users of electrical ground support equipment (EGSE) which is employed by the European industry during satellite construction.

The ETOL handbooks and associated documentation are readily available to assist users and potential users of ETOL. The handbooks present information in the form of user guides which are grouped into general purpose and specific application categories.

In addition, the following articles are also relevant to the contents of the book and these have not been directly referenced.

C. Green & B. Melton: ETOL development stimulates new approach for space software, *ESA Bulletin*, Number 39, August 1984.
G. Oppenhauser: ESA's first data-relay satellite experiment, *ESA Bulletin*, Number 47, August 1986.
J. T. Garner & G. S. Sims: Testing ground stations, *Space*, **2**, Number 24, December 1986.
J. T. Garner: Concepts of EGSE, *Journal of B.I.S.*, **40**, March 1987.
J. C. Bebruyn & C. S. Banks: REPPRE-REPSIM-REPSTA programs for evaluat-

ing the availability and maintenance of space systems, *ESA Journal*, **11**, Number 3, March 1987.

J. T. Garner & M. Jones: Rationalizing satellite control, *Space*, **3**, Number 3, July 1987.

J. T. Garner & M. Jones: ETOL and ESA communication satellite programmes, *Journal of B.I.S.*, **40**, September 1987.

R. D. Andressen & W. Nellessen: The EURECA concept and its importance in preparing for the Columbus programme, *ESA Bulletin*, Number 52, November 1987.

J. T. Garner, J. Sturcbecher, & M. Jones: Remote operation of communication satellites, *ESA Journal* **11**, Number 3, January 1988.

J. T. Garner: The ESA communication satellite monitoring facilities, *ESA Journal*, **14**, Number 1, March 1990.

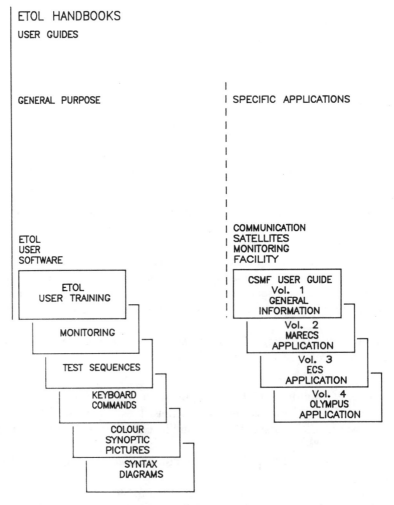

Fig. 1.

Appendix
ETOL applications

The on-line software modules MONITOR, TVPICT, and SEQUENCER function in unison, utilizing the ETOL database to enable received telemetry data to be processed and displayed. The results of processing the telemetry data and subsequent comparison with the data in the monitor tables are sent dynamically to the colour synoptic displays. In addition, the results of automatic sequences may be directed to sequence windows in colour synoptic displays. This appendix describes the structure and operation of the database.

A.1 CONTROL AND DISPLAY FACILITIES

The ETOL facilities enable the system executive control to be maintained, and a user to see the results of database operations.

A.1.1 Control

The control function is provided by the workstation VDU keyboards which allow ETOL commands to be generated. These commands fall into specific functions associated with the operation of the ETOL database, namely:

- ETOL monitoring control commands
- ETOL telecommand control commands
- ETOL test sequence control commands
- ETOL output device control commands
- ETOL system control commands.

Examples of these commands are given in Table A.1.

A.1.2 Display

The display feature can be provided by visual display units (VDUs), colour monitors, line printers, and graphical plotters. VDUs and associated keyboards are used to provide an active man–machine interface for spacecraft operations. Fig. A.1 is given as an example of a VDU output.

Table A.1 — ETOL keyboard commands

Function	Command
(1) ETOL Monitoring control	EM — Enable monitoring IM — Inhibit monitoring IN — Initialize monitoring MS — Display once a monitor section status UL — Update analogue parameter limit(s) DV — Display once current monitoring value(s) RV — Report repeatedly selected monitoring value(s)
(2) ETOL telecommand control	AS — telecommand address and synchronization word selection TC — Transmit telecommands
(3) ETOL test sequence control	EX — Execute test sequence(s) HD — Hold execution of test sequence(s) HL — Halt execution of test sequence(s) GO — Release held test sequence(s)
(4) ETOL output device control	CT — Insert comment into line-printer output SD — Select display of information shown on other VDU areas TV — Select colour synoptic picture for display on designated colour monitor unit
(5) ETOL system control	CF — Clear all flags EN — Enable background task LF — List status of flag(s) RF — Reset specified flag(s) SF — Set specified flag(s)

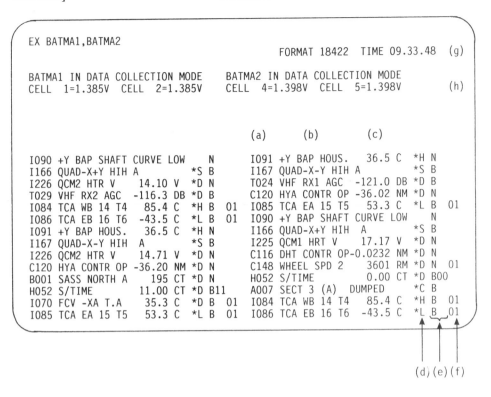

```
  EX BATMA1,BATMA2
                                    FORMAT 18422  TIME 09.33.48   (g)

  BATMA1 IN DATA COLLECTION MODE   BATMA2 IN DATA COLLECTION MODE
  CELL  1=1.385V  CELL  2=1.385V    CELL  4=1.398V  CELL  5=1.398V    (h)


                            (a)       (b)        (c)

  I090 +Y BAP SHAFT CURVE LOW    N    I091 +Y BAP HOUS.   36.5 C  *H N
  I166 QUAD-X+Y HIH A        *S  B    I167 QUAD-X-Y HIH A         *S B
  I226 QCM2 HTR V    14.10 V *D  N    T024 VHF RX1 AGC  -121.0 DB *D B
  T029 VHF RX2 AGC  -116.3 DB *D B    C120 HYA CONTR OP -36.02 NM *D N
  I084 TCA WB 14 T4   85.4 C *H B 01  I085 TCA EA 15 T5   53.3 C  *L B 01
  I086 TCA EB 16 T6  -43.5 C *L B 01  I090 +Y BAP SHAFT CURVE LOW     N
  I091 +Y BAP HOUS.   36.5 C *H N     I166 QUAD-X+Y HIH   A       *S B
  I167 QUAD-X-Y HIH A        *S B     I225 QCM1 HRT V    17.17 V  *D N
  I226 QCM2 HTR V    14.71 V *D N     C116 DHT CONTR OP-0.0232 NM *D N
  C120 HYA CONTR OP -36.20 NM *D N    C148 WHEEL SPD 2    3601 RM *D N 01
  B001 SASS NORTH A    195 CT *D N    H052 S/TIME         0.00 CT *D B00
  H052 S/TIME       11.00 CT *D B11   A007 SECT 3 (A)   DUMPED    *C B
  I070 FCV -XA T.A    35.3 C *D B 01  I084 TCA WB 14 T4   85.4 C  *H B 01
  I085 TCA EA 15 T5   53.3 C *L B 01  I086 TCA EB 16 T6  -43.5 C  *L B 01
```

                                                          (d) (e) (f)

Fig. A.1 — VDU screen normal display (example) (courtesy of ESA).

Interpretation of Fig. A.1.
(a) Monitor parameter identifier (alphanumeric code).
(b) Monitor parameter label
(c) Monitor parameter value/condition
(d) Reason for monitor output message:
- Analogue parameters
   *H: above HIGH limit defined in monitor tables
   *L: below LOW limit defined in monitor tables
   *D: outside DELTA limit defined in monitor tables
   **: outside DANGER limit defined in monitor tables
   *F: parameter output due to use of ETOL command RV
       or DV
- Digital parameters
   *S: status of parameters not conforming to conditions specified in monitor tables
   *C: change in status of parameter conditions specified in monitor tables
(e) Telemetry data source
- Format
       N=normal format
       L=launch format
       B=both format
- Number
   Indicates the processed word number of a super-commutated parameter sample.
(f) Limit set utilized to give output message.
(g) Time in GMT (Zulu) when format was processed.
(h) Information from operating test sequences.

A.2 DATABASE

The ETOL database comprises:
- Colour synoptic pictures
- Monitor tables
- Test sequences.

A.2.1 Colour synoptic pictures

The colour synoptic pictures which display telemetered parameters can be constructed in the form of circuit diagrams, block diagrams, and/or in alphanumeric form. The pictures can also be constructed to be a combination of these types of display.

A library of pictures can be constructed for each satellite model. For ESA communication satellites, a typical library comprises approximately fifty synoptics. *Note:* Examples are given in Figs A.2–A.9 (in colour section).

Fig. A.2 shows a colour synoptic picture which is a snapshot that gives an overview of the conditions and status of ECS-5. The colour synoptic library provides several such pictures for each on board subsystem, and these pictures can be constructed to give a zoom into the boxes shown in Fig. A.2. For example, Figs A.3 and A.4 are zooms into the upper part of Fig. A.2 and relate to the on board power supply subsystem.

ETOL, besides presenting data in the form of a block or circuit diagram, also allows data to be displayed as an alphanumeric snapshot. Fig. A.5 is an example of this, where the OLYMPUS-1 satellite combined propulsion system temperatures are displayed in alphanumeric form. This system provides and controls the on board fuels for both the AOCS and the liquid fueled ABM.

In all of these colour synoptic pictures, a green background indicates that the value(s) is within limits, and a red background indicates that high or low limits have been exceeded. The figure(s) have been chosen to demonstrate the operation of this facility, consequently some perfectly acceptable values have been deliberately displayed on a red background.

Figs A.6, A.7 and A.8 (see colour section) demonstrate how synoptic pictures can be employed to display information that is not a snapshot or a direct part of inter-segment operations. In these cases, the synoptics can be employed to display procedural information.

The colour synoptic pictures are generated by using the ETOL software module PICTGEN.

A.2.2 Monitor tables

Monitor tables are constructed to define all the characteristics associated with spacecraft telemetry parameters. In addition, monitor tables allow telecommand orders to be constructed, and checked against expected predefined values. This provides the closed-loop control which can be employed by the telemetry and telecommand interfaces between space and ground segments. Monitor tables are normally generated as an off-line task using the ETOL monitor table generation program (MTGP) via an interactive dialogue. The on-line program MONITOR utilizes the monitor tables when processing telemetry data.

The following explanation of these tables is given in a general form to portray the most pertinent aspects. Hypothetical examples of telemetry and telecommand monitor table outputs from the MTGP are given in Fig(s) A.10 and A.11. More details can be obtained from associated documents referenced in the *Bibliography*.

Telemetry Monitor Tables

These tables have five prime fields, with each field consisting of several areas. The information needed for these areas in some instances is supplied by the MTGP after an interrogative dialogue has been undertaken. In other cases,the information needs to be entered directly by the user. The user interacts with the MTGP at workstations via VDUs and associated keyboards to construct the monitor tables in accordance with the constraints of their definitions.

The five prime fields are:

Field 1: Parameter definition.
Field 2: Options.
Field 3: Sampling.
Field 4: Applicable to parameter type (analogue or digital).
Field 5: Calibration curve.

A more detailed explanation of the fields associated with telemetry monitor tables is given on page 154 and should be used in conjunction with Fig A.10.

Telecommand monitor tables

These tables require information concerning the following four areas:

- Telecommand definition
- Telecommand type
- Command address
- Telecommand condition

Again, interactive processes prevail between the MTGP and the user at a workstation (VDU and keyboard). A more detailed explanation of the telecommand monitor table fields are given on p. 155 in conjunction with Fig A.11.

A.2.3   Test sequences

A test sequence is a series of ETOL statements which constitute a program to be executed under the control of the on-line module SEQUENCER. Test sequences may be written by a satellite engineer in the high level language ETOL to enable automated tests to be performed. Once verified they become part of the ETOL database and function in unison with monitor tables and colour synoptic pictures.

Example

To demonstrate how spacecraft operational activities can be automated, the following hypothetical example is given. It explains the principles of test sequence operation and how the correlation between the ETOL database elements can be achieved.

The colour synoptic picture Fig A.9 in colour section together with the telecommand and telemetry parameters given in the monitor tables Figs A.10 and A.11 are used to demonstrate the way in which the ETOL facilities may be employed.

```
Parameter: E 1      Label: UN-A-STATUS
Paramtype: DIGITAL
OPTIONS  :                                    Linked to command: YES
           Special Processing: 0    Status texts: NO   Conditions: NO

SAMPLING :
           Supercomm.: NO
           Channels Location:   Chann. 1 Frame:  4 Loc:  17   Mask: £1

Parameter: E 2      Label: UN-B-STATUS
Paramtype: DIGITAL
OPTIONS  :                                    Linked to command: YES
           Special Processing: 0    Status texts: NO   Conditions: NO

SAMPLING : ex fr no:   0
           Supercomm.: No
           Channels Location:   Chann. 1  Frame:  4 Loc:  17   Mask: £2

Parameter: E 3      Label: UN-C-HEATER
Paramtype: DIGITAL
OPTIONS  :                                    Linked to command: YES
           Special Processing: 0    Status texts: YES  Conditions: NO

SAMPLING :
           Supercomm.: NO
           Channels Location:   Chann. 1  Frame:  4 Loc:  17   Mask: £8

Ststext  : Status No        From        To        Status Text
              1              0          0         DISABL
              2              1          1         ENABLE

Parameter: E 4      Label: UN-C-TEMP
Paramtype: ANALOGUE              Format: REAL    Scale: 2    Units: C
OPTIONS  :              Calibration: 1    Apply Danger Monitor: NO
           Special Processing: 0    Limits. YES     Conditions: NO

SAMPLING :
           Supercomm.:  NO
           Channels Location:   Chann. 1  Frame: 7 Loc: 13    Mask: £FF

Limits   : Deltalimit: YES  Counter: NO   Limit Selection: NO
           Set No.       Low          High          Delta
              1         -2.00         28.00          0.50

           Calibration Curve No: 1  Format: REAL  Scale: 2

              Point no        X-Coord        Y-Coord
                 1            (29            40.00  )
                 2            (66            30.00  )
                 3            (112           20.00  )
                 4            (156           10.00  )
                 5            (204            0.00  )
                 6            (248           -7.00  )
```

Fig. A.10 — Telemetry monitor table parameters (courtesy of ESA).

Explanation of telemetry monitor tables — (Fig. A.10)

FIELD 1	: MONITOR PARAMETER DEFINITION
Parameter	: Alphanumeric identifier

The following limitations are applicable: Alphabetic A–Y, Numeric 0–255. Normally, the alphabetic identifier is aligned to the associated subsystem. For example, P123 for a Power subsystem parameter is pertinent, in this case.

Example 1 E1

Label	: Descriptive identifier

examples: UNit A STATUS   (parameter E1)
                UNit C TEMPerature (parameter E4)

Paramtype	: DIGITAL or ANALOGUE

end of field for digital parameters for analogue parameters field 1 includes:

Format	: REAL/INTEGER/HEXADECIMAL
Scale	: Range of value definition

Selection from the following table:

Range of values	Scale code
− 0.999 to + 0.9999	0
− 9.999 to + 9.999	1
− 99.99 to + 99.99	2*
−999.9 to +999.9	3
−32768 to +32767	4

*Definition selected for parameter E4

Units	: Engineering units identifier

For parameter E4, degrees Celsius

FIELD 2	OPTIONS

Dependent on parameter type.
The most pertinent points for parameters E1 and E4 follow:
: E1 (Digital) It will be seen in the example that the monitoring of this parameter is linked to telecommand(s) transmission. No special processing, status texts, or conditions are applicable.
: E4 (analogue) It will be seen in the example that one calibration curve is applicable and that limits are to be applied to the monitored values. No danger limit monitoring, special processing, or conditions are applicable.

FIELD 3	SAMPLING

This field defines the location of the telemetry channel within the telemetry format.

Supercomm	: Super commutated channel YES or NO.

In the examples NO
therefore, one channel only

Channels Location	: Chann.1

Located within

Frame	: Number 7 for E4

Number 4 for E1

Loc	: word location within specified telemetry frame.

Word 17 for E1
Word 13 for E4

Mask	: Bits applicable to channel

defined in hexadecimal (£)
E1 bit $2°$(bit no. 0)=1
E4 all eight bits=FF

FIELD 4	APPLICABLE TO PARAMETER TYPE

For analogue parameters a limit definition is applicable.

Limits	: As defined in this field

For parameter E4 one set of limits is defined.
For Digital parameter no limits applicable, but status text is.

Ststext	: Status text

Parameter E1 — NO special definition required as indicated in FIELD 2, therefore the following text definitions will be used.
                0=OFF
                1=ON
Parameter E3 — YES special definition required as indicated in FIELD 2 and defined in this field.
                0=DISABLE
                1=ENABLE

FIELD 5	CALIBRATION CURVES

For the example of this appendix there is only one analogue parameter and a single calibration curve within this field.
The curve is defined with
        Telemetry values: X co-ordinates
        Engineering values: Y co-ordinates
The definitions format: and scale: are as explained in FIELD 2.

Note: When outputs are made from MTGP, FIELD 5 is displayed before FIELD 1.

```
Telecommand: Z 10    Label: UN-A-PWR-ON
Tctype: D1    '                    Command Address: £C0
           Command Status: ENABLED       Expected Value Chain(s): YES

Tc Condition 1   :   IF 1
                     Then Expect
                 1   Param E  1    = Cmnd data:  NO    Value: 1


Telecommand: Z 11    Label: UN-A-PWR-OFF
Tctype: D1                         Command Address: £68
           Command Status: ENABLED       Expected Value Chain(s):  YES

Tc Condition 1   :   IF 1
                     Then Expect
                 1   Param E  1    = Cmnd data:  NO    Value: 0


Telecommand: Z 12    Label: UN-B-PWR-ON
Tctype: D1                         Command Address: £C1
           Command Status: ENABLED       Expected Value Chain(s):  Yes

Tc Condition 1   :   IF 1
                     Then Expect
                 1   Param E  2    = Cmnd data:  NO    Value: 1


Telecommand: Z 13    Label: UN-B-PWR-OFF
Tctype: D1                         Command Address: £69
           Command Status: ENABLED       Expected Value Chain(s):  YES

Tc Condition 1   :   IF 1
                     Then Expect
                 1   Param E  2    = Cmnd data:  NO    Value: 0


Telecommand: Z 14    Label: UN-C-HTR-EN
Tctype: D1                         Command Address: £C2
           Command Status: ENABLED       Expected Value Chain(s):  YES

Tc Condition 1   :   IF 1
                     Then Expect
                 1   Param E 3     = Cmnd data:  NO    Value: 1


Telecommand: Z 15    Label: UN-C-HTR-DIS
Tctype: D1                         Command Address: £6A
           Command Status: ENABLED       Expected Value Chain(s):  YES

Tc Condition 1   :   IF 1
                     Then Expect
                 1   Param E 3     = Cmnd data:  NO    Value: 0
```

Fig. A.11 — Telecommand monitor table parameters (courtesy of ESA).

Explanation of telecommand monitor tables — (Fig A.11)

TELECOMMAND PARAMETER DEFINITION

Parameter
: Similar to telemetry tables except that all telecommands are automatically assigned the alphabetic identifier 'Z', and numerical range is 0–999

Label
: Descriptive identifier
example: UNit-A-PoWeR-ON for telecommand Z10

Tctype
: Telecommand type, for example direct mode 1 (on/off)

Command address
: Data word 1 value in hexadecimal (L) C0 (telecommand Z10)

Command status
: For all the examples telecommands are ENABLED.
i.e. not on the inhibited command list.

Expected value chain(s)
: For all examples the telecommand conditions are linked to a single chain which is defined in the next field.

Tc condition 1
: For the examples there is only one condition which is expected to result from successful command executions. For each telecommand, the monitoring of this is defined as the change expected in the associated telemetry parameter.
In the examples, a value/status change from $0 \rightarrow 1$ or $1 \rightarrow 0$ for telemetry parameters E1, E2, E3
For these telecommand examples the command data word need not be checked only the result of the telecommand execution as defined in the expected parameter value.

Before proceeding further, the following points regarding the normal operation of ETOL need to be considered.

- If a test sequence is operating and ETOL processing stops, possibly owing to the loss of the telemetry signal, the test sequence will halt and a message will be sent automatically to the display devices.
- A test sequence restart will require manual actions.
- An abnormal condition requiring the attention of a satellite engineer can result in a programmed halt.

The definition of the elements shown in Fig. A.9 are as follows:

Unit A — Power supply (prime)
Unit B — Power supply (redundant)
Unit C — On-board unit which requires temperature control
SW1   — Enables unit A to be powered ON/OFF
SW2   — Enables unit B to be powered ON/OFF
SW3   — Enables heater current to unit C to be switched ON/OFF

Fig. A.9 depicts the results of ETOL operations when the following conditions exist:

SW1 closed: E1 ON
             Green indicator illuminated on unit A
SW2 open: E2 OFF
             Red indicator illuminated on unit B
             SW3 closed: Heater enabled
                            Green indicator illuminated on unit C
Unit C: Within high/low limits (15°–25°C) — blue
         Out of limits — red

The indentifier of this colour synoptic is P101, and this will be found on the bottom left-hand side of the picture.

This example demonstrates how the ETOL user software monitors and controls the temperature of unit C; parameter E4 in the telemetry monitor tables (Fig A.10). The value is displayed below the appropriate block on the synoptic picture (Fig A.9). Any necessary manual actions could then be taken which would result in tempera-ture corrections for unit C.

For this example, the actions are performed automatically by a test sequence. Before writing the test sequence it is necessary to define the requirements. These are as defined by the following summarized operational procedure:

Unit C heater to be switched ON/OFF (SW3) when LOW/HIGH limits respec-tively are exceeded. The heater can remain ON or OFF until the opposite limit is exceeded, when appropriate action shall then be taken. The delta limit can be exceeded without any special actions being undertaken. It should be noted that whilst the temperature of unit C can be raised by means of a heater, there is no active device available for reducing the temperature. Cooling relies on heat conduction inherent in the design of the satellite structure.

Table A.2 — User requirements

Number	Description
1	NOMINAL CONDITIONS Unit C is operational: heater disabled. Units A and B are off. Temperature of unit C at nominal value, i.e. 15°C approximately.
2 (a)	INITIALIZATION — MANUAL OPERATIONS Verify that telemetry signal is being received and decommutated. Selection of unit A or B power supply to unit C heater is to be undertaken and action verified. (Send TC Z10 or Z12).
(b)	After requirement (a) has been completed successfully then unit C heater to be enabled and action verified. (Send TC Z14).
(c)	After requirement (b) has been completed successfully then requirement 3 to be undertaken automatically (start ETOL test sequence UNCHTR).
Note:	After heater switch on has been implemented then temperature of unit C will increase. Without the use of heater unit C temperature will reduce and exceed low limit constantly.
3	AUTOMATIC OPERATIONS Unit C temperature is to be maintained within the high/low limits defined in the monitor tables and the temperature value is to be displayed as part of a colour synoptic picture on an output device. The delta limit can be exceeded without any corrective actions being undertaken.
4	TERMINATION — MANUAL OPERATIONS End automatic operations, disable unit C heater and switch off unit A and B power supplies. All actions to be verified. (Halt ETOL test sequence and send Z11, Z13, and Z15.)

Table A.2 provides a definition of the user requirements, and is complemented by Fig. A.12 which is a flow chart specifying how these requirements will be achieved. Fig A.13 results from compiling a test sequence and printing a list; part of the activities associated with the operation of the ETOL test sequence compiler module. It is an exact definition of the test sequence which will satisfy the software requirements given in Table A.2 and Fig. A.12. A few explanatory notes regarding this listing are given below:

- The vertical line of numbers on the extreme left-hand side refer to the executable ETOL test sequence statements.
- Statements 6,8,19 refer to messages which should be displayed in P101 (Fig. A.9).
- Statements 15,24 define the message which should be displayed on the VDU (Device A2 row 4 column 2) of the test conductor console.

INITIALIZATION — ETOL keyboard commands — (manual)
(a) TC Z10   — or Z12 ⎫ Telecommand transmissions initiated and
(b) TC Z14   —   ⎬ results verified (manual) on display devices.
(c) EX UNCHTR   — EXecute (start)
                    Test sequence
                    named UNCHTR

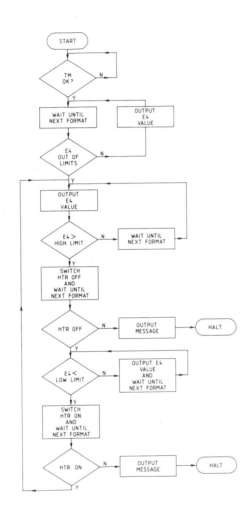

TERMINATION
(a) HL UNCHTR — HaLt
                Test Sequence
(b) TC Z11,Z13,Z15 — Telecommand transmissions initiated and
                    Results verified (manual) on display devices
The HL command can be sent and actioned at any time.

Fig. A.12 — Software flow chart (courtesy of ESA).

```
SEQ UNCHTR                                   Sequence name UNit C HeaTeR
COMMAND HTREN = Z14;                         Z14 HeaTeR ENable
COMMAND HTRDIS = Z15;                        Z15 HeaTeR DISable

      L1   :
0          IF TMSYNC=TRUE;
1            THEN GOTO L2;                    Check TM sync
2            ELSE GOTO L1;
      L2   :
3          WAIT FRAME 0;                      Wait for start of format
4          IF E4=NOGO;                        Check E4 status (exeeds limits ?)
5            THEN GOTO L3;                     Not O.K.
             ELSE BEGIN;                       O.K.
6              OUTPUT TV(P101,1) E4;          Display E4 value in synoptic picture
7              GOTO L2;
             FINISH;
      L3:
8          OUTPUT TV(P101,1) E4;              Display E4 value in synoptic picture
9          IF E4 > 28;                        Check high limit
10           THEN SEND HTRDIS;                If high switch heater off (Z15)
             ELSE BEGIN;
11             WAIT FRAME 0;                  No, wait start of format
12             GOTO L3;
             FINISH;
13         WAIT FRAME 0;                      Wait start of format
14         IF E3 = 1;                          Is heater off?
             THEN BEGIN;
15             OUTPUT DA2(R4,C2)"TC Z15 FAIL";  No, then output message
16             STOP;                                Halt test sequence
             FINISH;
      L4:
17         IF E4 < -2;                         Check low limit
18           THEN SEND HTREN;                 Yes, switch heater on (Z14)
             ELSE BEGIN;
19         OUTPUT TV(P101,1) E4;              Display E4 value in synoptic picture
20             WAIT FRAME 0;                  No, wait next format and try again
21             GOTO L4;
             FINISH;
22         WAIT FRAME 0;
23         IF E3 = 0;                          Is heater on?
             THEN BEGIN;
24             OUTPUT DA2 (R4,C2)"TC Z14 FAIL";  No, then output message
25             STOP;                                Halt test sequence
             FINISH;
26         GOTO L3;                           Return to high limit check
27         END;
```

Fig. A.13 — Test sequence listing (courtesy of ESA).

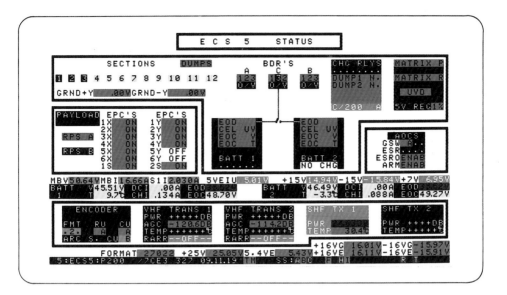

Fig. A.2 — Overall status/condition of ECS-5. The top block of the picture relates to the power supply subsystem. Blocks associated with the communications payload and the attitude and orbit control subsystem (AOCS) can easily be identified. The bottom block relates to the telemetry, tracking, and telecommand (TTC) subsystem. The information immediately above the lower block is general information associated with the power supply subsystem (courtesy of ESA).

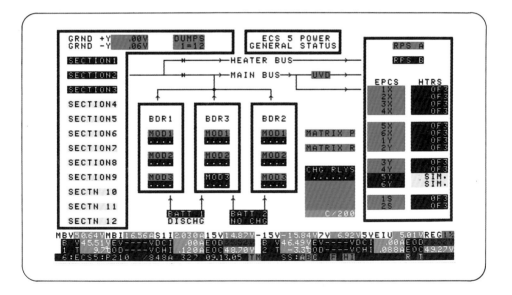

Fig. A.3 — General status of ECS-5 power supply subsystem. The block on the left-hand side indicates which sections of the solar array are delivering power, i.e. those on a yellow background. The block on the right-hand side indicates which sections of the payload repeaters and thermal heaters are powered. Between these blocks the conditions and status of the battery discharge regulators (BDR) are indicated (courtesy of ESA).

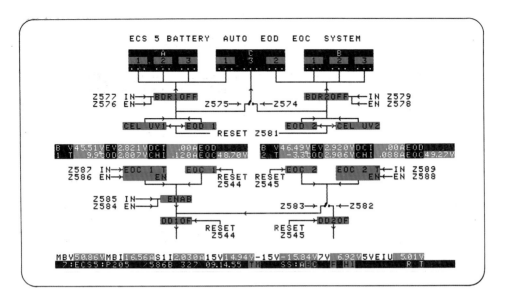

Fig. A.4 — ECS-5 battery charge/discharge system. This picture displays the condition and status of parameters associated with the semi-automatic operation of the power supply's BDRs. The telecommands which cause status changes are indicated by the alphanumeric Z parameter (courtesy of ESA).

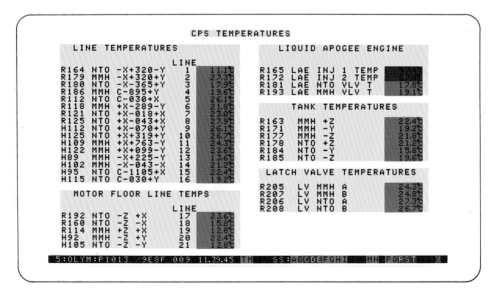

Fig. A.5 — OLYMPUS-1 combined propulsion system temperatures. An alphanumeric snapshot which displays temperature values associated with the combine propulsion system (CPS) (courtesy of ESA).

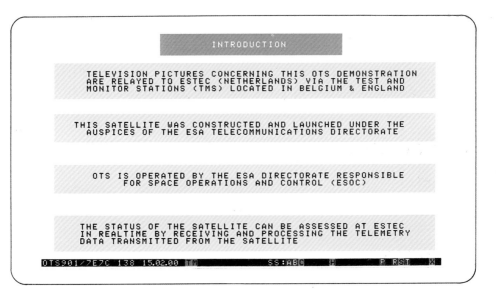

Fig. A.6 — Text information display. A demonstration of how information that is not derived from telemetry signals can be displayed (courtesy of ESA).

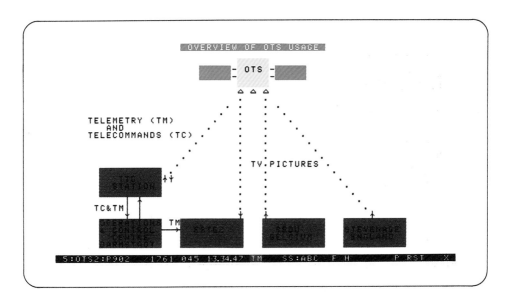

Fig. A.7 — Information displayed in synoptic form. This picture also portrays information which is not directly associated with active inter-segment operations. It follows Fig. A.6 in a sequence that can be executed by an ETOL test sequence (courtesy of ESA).

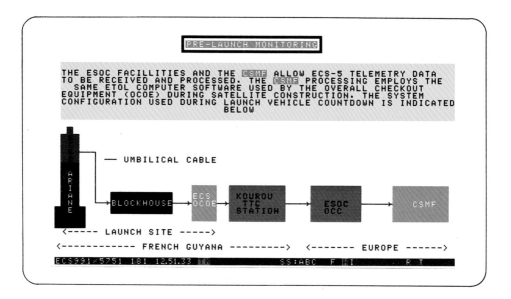

Fig. A.8 — Synoptic and text display. A combination of the possibilities which have been shown in Fig. A.6 and Fig. A.7 (courtesy of ESA).

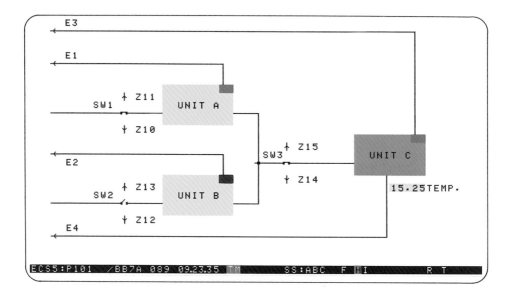

Fig. A.9 — Synoptic example A demonstration of how ETOL database elements can be operated in unison (refer to the explanations given in section A.2.3) (courtesy of ESA).

Index